ADVANCES IN LASER AND OPTICS RESEARCH
VOLUME 1

ADVANCES IN LASER AND OPTICS RESEARCH
VOLUME 1

WILLIAM T. LAVWORTH
EDITOR

Nova Science Publishers, Inc.
Huntington, New York

Senior Editors: Susan Boriotti and Donna Dennis
Coordinating Editor: Tatiana Shohov
Office Manager: Annette Hellinger
Graphics: Wanda Serrano
Book Production: Matthew Kozlowski, Jonathan Rose and Jennifer Vogt
Circulation: Cathy DeGregory, Ave Maria Gonzalez, Ron Hedges and Andre Tillman

Library of Congress Cataloging-in-Publication Data
Available upon request.

ISBN 1-59033-056-0

Copyright © 2002 by Nova Science Publishers, Inc.
227 Main Street, Suite 100
Huntington, New York 11743
Tele. 631-424-6682   Fax 631-425-5933
e-mail: Novascience@earthlink.net
Web Site: http://www.nexusworld.com/nova

Printed in the United States of America

Contents

Preface

It is expected that ongoing advances in optics will revolutionize the 21st century as they began doing in the last quarter of the 20th. Such fields as communications, materials science, computing and medicine are leaping forward based on developments in optics. This new series presents leading edge research on optics and lasers from researchers spanning the globe.

High Power Semiconductor Lasers: Different Systems

Salvador Guel-Sandoval
Instituto de Investigación en Comunicación Optica
U.A.S.L.P.
Av. Karakorum 1432, Lomas 4a. Sección,
San Luis Potosí, S.L.P. 78000, UASLP
sguel@cactus.iico.uaslp.mx

Abstract

An important limitation in semiconductor lasers is their non-capability to produce a high power energy amount, simultaneously with a high degree of spatial coherence. This is: if high power is desired, one must increase the lasing active volume; however this also increases the number of spatial modes thus diminishing the spatial coherence, and vice-versa. Accordingly a compromise must be established between both quantities if an optimal functioning of such devices is desired. In this work, the different methods proposed to overcome such difficulty, since the semiconductor laser invention to the present decades are discussed.

1 Introduction

Since its invention in 1962, [1,2,3,4] the GaAs semiconductor laser has become a reliable and efficient miniature source of coherent light, calling immediately the attention of scientists and engineers to its multiple applications. Since then, semiconductor laser technology has registered tremendous advances that are continuing thorough the 2000's. The key to the expanding capabilities of semiconductor diode lasers has been the development of new structures, compounds and materials growth techniques, in the continuing search for higher reliability and efficiency, better control and higher power operation [5].

A key area of research concerns two of the basic characteristics of a diode laser: the power that can be obtained and the spatial coherence, which determines the capability of the laser light to be focused. While many applications of diode lasers, such as optical pumping of solid state lasers, infrared laser illumination, laser soldering, eye surgery and others depend only on the level of power that can be obtained from the laser, many other as optical data storage, telecommunications via single mode fibers, satellite communication, printing, etc. require power and a high degree of spatial coherence.

Obtaining both high power and high spatial coherence simultaneously in a diode laser represents a difficult technical challenge, caused primarily by the multimode behavior of the laser at high power levels. This need has resulted in much research work in the search for a laser design that includes both characteristics.

Modes in a laser diode can be longitudinal or spatial. Longitudinal modes are illustrated in figure 1. Physically each longitudinal mode is a standing wave of light, created by the overlap of two traveling waves that are moving in opposite directions. Spatial modes, in either the lateral or transverse directions, are depicted in figure and are represented as wavefronts perpendicular to the beam propagation. Strictly speaking, transverse modes of light, electric (TE) or magnetic (TE), (depending on which field component, vector $\mathbf{E}$ or vector $\mathbf{H}$, is perpendicular to the direction of beam propagation) in a given medium, are mathematical solutions to the wave equation in that medium [6]. These solutions physically represent waves, being these waves often referred as TEM_{nm} optical waves where n and m stand for the order [7]. This notation have been used in Fig. 2.

Figure 3 shows the normal orientation of longitudinal, transverse and lateral modes for a semiconductor laser. Both longitudinal and transverse modes have discrete light energy distributions, whose number or order is determined in the first case by how many times an integer number of half-wavelengths can fit inside the laser cavity and in the second by the allowed values of the propagation. Generally the higher the number of longitudinal or spatial modes, the smaller de degree of temporal or spatial coherence.

Multiple spatial modes, either transversal or lateral, result in incoherence as each will have a well defined and different phase. The problem of multiple transverse modes is eliminated in heterojunction devices by the optical confinement that is offered by the lower index materials surrounding the gain region. Multiple lateral modes are traditionally avoided by decreasing the width of the gain stripe to ~3 mm, where only one lateral mode is pumped. This method, however, has the disadvantage that the power density increases to levels at which catastrophic facet damage can occur.

From these considerations, the importance of the search for a method to obtain not only high power from a semiconductor laser but also high spatial coherence has been of the interest of many people. In the following paragraphs the most representative advances in this matter, will be mentioned.

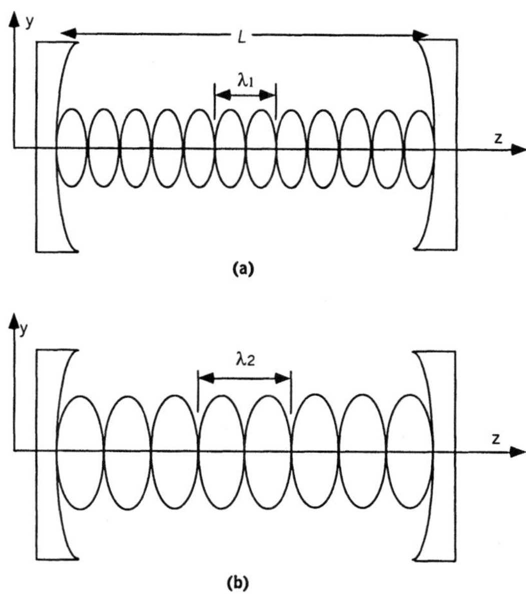

Figure 1: Illustrating the longitudinal mode concept. If the laser cavity length L is equal to an integer number of half-wavelengths, a stationary wave-train is formed since waves traveling in opposite directions interfere constructively. The figure shows two longitudinal modes corresponding to λ_1 and λ_2. In (b) the wave amplitude is not necessarily larger than in (a).

2 The evolution of high power coherent laser technology

During the past decades considerable efforts have been made to obtain reliable high power coherent semiconductor lasers. Even though some methods have given satisfactory results, the search continues for better and more efficient laser designs.

The first reported lasers worked in very low temperature conditions (77°K), to better tolerate the resistive heating produced by the high drive current [1-4]. One of the most significant advances in semiconductor laser technology was introduced by Hayashi *et al* and other working groups at the late 60's and early 70's [8]. They demonstrated that lasers containing GaAs-Al$_x$Ga$_{1-x}$As, liquid phase epitaxy (LPE) grown structures could operate continuously at room temperature. The improved carrier and optical confinement achieved by this method lowered the threshold current density (J_{th}) by several orders of magnitude up to ~1.6 kA/cm^2 in uncoated Fabry-Perot diodes 560 μm long. Power levels of 20 mW were reported, but the major contribution of this work was to demonstrate the possibility of obtaining lasers diodes working under room temperature conditions. (Note: terminology and concepts referring to the laser beam quality characterization such as: far field, degree of coherence, fringes visibility, beam divergence, full width at the half of the maximum (FWHM), diffraction limited, quantum efficiency etc., will be discussed in the appendix A of this work).

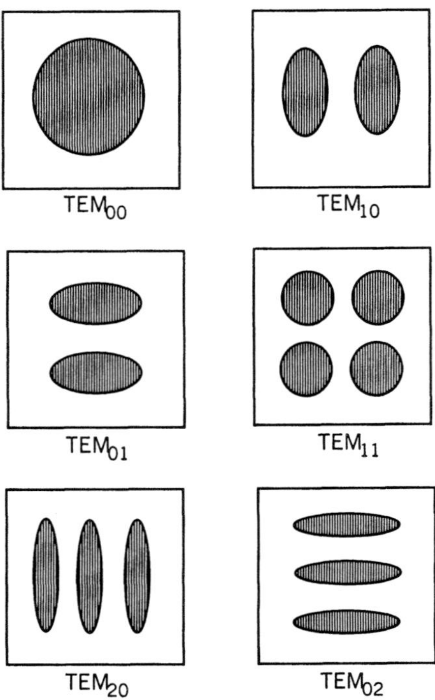

Figure 2: Illustrating the spatial mode concept. As light bounces back and forth inside the cavity only in fixed directions, it forms stationary spatial wavefronts or modes. The figure shows a laser stable resonator transverse section, when lasing in different spatial modes. The intensity patterns of the spatial modes are indicated by numbers. The left number represents the minima in the horizontal direction; the right number, the minima in the vertical direction. TEM$_{00}$ would be the fundamental order mode.

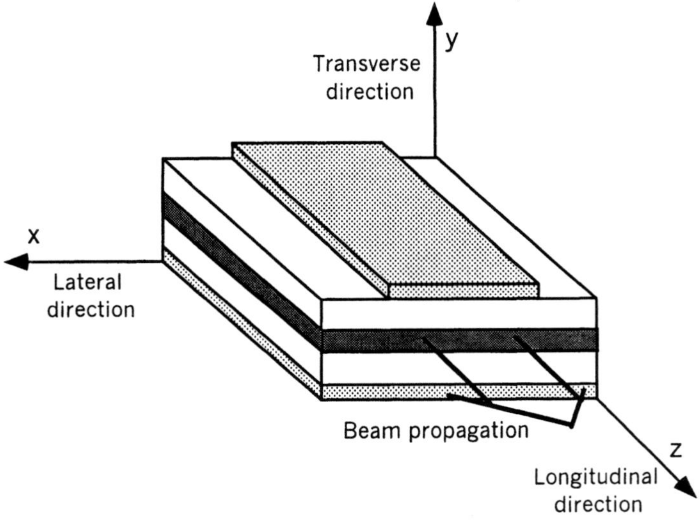

Figure 3: Coordinate axis system used to define lateral, transverse and longitudinal (beam propagation) directions, represented by x, y, z, axis respectively.

During the late 70's. Dupuis *et al*, demonstrated improved double heterostructures grown by metalorganic chemical vapor deposition (MOCVD) [9]. Threshold current density values of 590 A/cm^2, were achieved who were attributed to the high quality of the GaAs and Al$_x$Ga$_{1-x}$As, layers materials produced by MOCVD process. Later, by using multiple very thin active layer structures (the concept of quantum well (QW) was introduced for the first time), this same group obtained external quantum efficiencies (η_d) as high as 85%. However, the threshold current density was increased [10]. Further experiments, using more sophisticated laser structures and crystal growth methods, resulted in even lower values of J_{th} than the previously reported.

In the earlies 80's the graded-index separate-confinement heterostructure (GRIN-SCH) concept was introduced. Here the quantum well active region was surrounded by either a continuously [11] or step graded [12] confining region, and this further J_{th} to ~250 A/cm^2. These low threshold current designs significantly reduced Joule heating in the laser and opened the door to high power operation.

2.1 Phase Locked Multichannel Arrays

One of the more successful approaches to high power is the phase-locked multichannel array initiated by Scifres *et al* en 1978 [13]. With this technique, shown schematically in figure 4, output beams with less than 2° of divergence were achieved by coupling radiation between several adjacent narrow stripe lasers. The contacts used in this early work were typically 3.5 µm wide and stripes were separated 8 µm center top center. In this first work a power 60 mW per facet was reported, with a quantum efficiency greater than 25 % per facet.

Improvements were achieved by using a variation of this method, the so called: coupled multiple stripe (CMS) or more recently Y-junction array illustrated in figure 5 [14]. Reinforcement of in-phase waves was achieved by making them coincide in a common path, obtaining in this way up to 0.9 W/facet with a beam divergence of less than 4° in pulsed operation at room temperature. The threshold (I_{th}) was 400 mA, and a η_d of 65% was observed.

Even better results were obtained later by this same group by using a multiple quantum well (MQW) structure [15]. Devices this time emitted up to 2.1 W/facet, with peak pulsed (75 ns) output power and 400 mW/facet in continuous wave (CW), but no specific degree of coherence was reported. In this case MQW structures achieved more power since the lasing volume increased. Those devices were called: CMS quantum well injection laser array.

A further advance using this same technique was achieved when the first CW array devices were demonstrated to be capable of operation at power levels greater than 200 mW/facet, at 100 °C and 600 mW/facet at room temperature [16]. Threshold current I_{th} was 130 mA and the external quantum efficiency was ~60 %. The far field divergence was too broad this time (60° at FWHM), so the interest in this experiment was for achieving CW laser functioning at high power. No multiple quantum well structures were used this time.

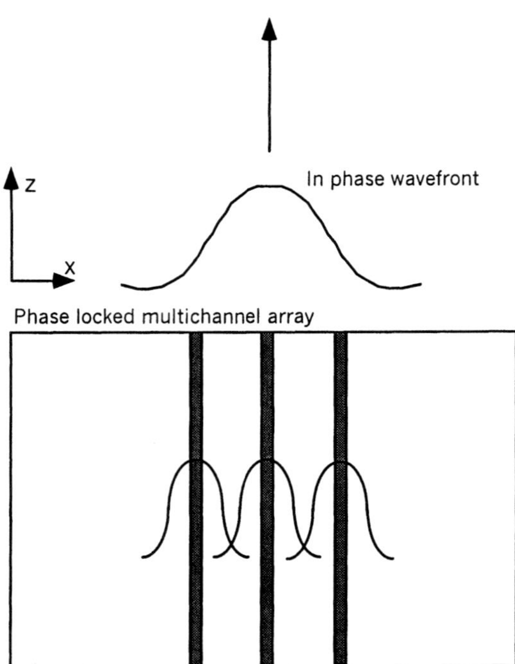

Figure 4: Schematic representation of a phase-locked multichannel array. By optically coupling the radiation between adjacent stripes, a high degree of phase-locking and lateral mode stability is achieved.

Ackley and Engelmann proposed an interesting variation to these schemes as shown in figure 6 [17, 18]. By enclosing the lasing gain medium between two higher index cladding regions, using buried structures, an antiguide was formed in the lateral direction (something like a negative waveguide). This led to a strong coupling between lasing stripe lasers through the leaky-modes propagation along the high index regions. Later with an improved model called channel-guide (CG) laser, outputs up to 2.8 W with a stable far field were measured using ten stripe structures. A unique feature of these negative-index-step guided devices was the fact that they showed both strong optical confinement and strong overall inter-element coupling.

However an undesirable characteristic of the phase-locked arrays was the formation of array modes due to the allowed wave coupling configurations between interelements. The two basic were: the in-phase (if there was no phase difference between adjacent elements) and the out-of-phase (if there was a π phase difference between adjacent elements).

By carefully calculating the separation between the anti-guiding regions, an array modes discrimination mechanism was introduced by D. Botez *et al*. Resonant lateral conditions were incorporated in the anti-guided array, thus increasing the capability of coupling between in-phase (resonant) modes, since near the resonance condition maximum discrimination against out-of-phase (non-resonant) modes was achieved [19]. These resonant optical waveguide (ROW) laser arrays were developed to the point that they were considered as practical sources of high coherent power [20]. Up to 2.1 W of power in single operation at diffraction limit (DL) and up to 3X the DL, has been reported with this method [21].

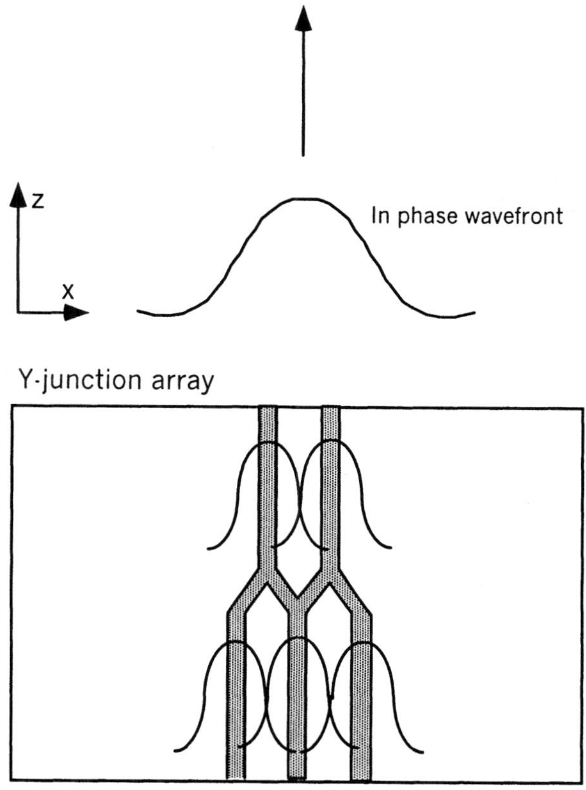

Figure 5: Illustration of the *Y*-junction coupler array. As in previous case, when the incident modal amplitudes are in-phase they add to excite a stable mode in the output waveguide.

As more sophisticated regrowth techniques became available so was the capability of improving modeling based on the original leaky-model idea. In 1990, T.S. Shiau *et al* from the Center for High Technology Materials at the University of New Mexico, reported for the first time, operation of a strained QW leaky-mode laser in which the antiguiding regions, to form the laser structure, were incorporated into the substrate by a procedure of growth-etch-regrowth technique using metallorganic chemical vapor deposition (MCCVD), thus simplifying the method already proposed that consisted of several more steps. Reported data from this experiment were: 172 mW of pulsed optical output power per facet and a far field of 1.6X the DL at FWHM [22]. Later this technique was successfully implemented by D. Meyhus *et al*, to produce up to 1 W under CW conditions, and up to 10 W under short-pulse (50 ns) operation with a far-field between 2-3 times the DL, using typically 20 elements arrays [23].

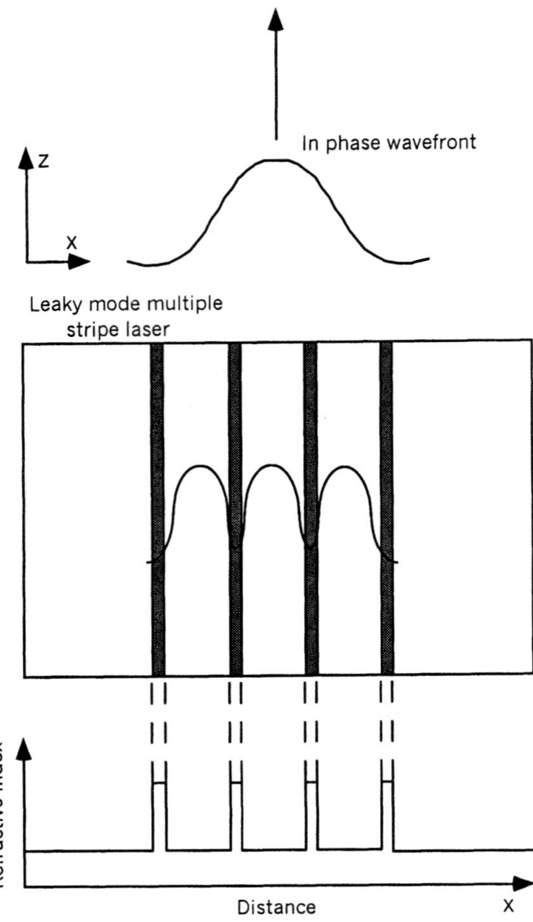

Figure 6: Illustrating the antiguide coupled array concept. Gain is in low index regions. Coupling is made through leaky modes propagating along the high index regions.

A different method for array modes discrimination in coupled laser arrays was based on the Talbot effect. According to the Talbot effect, an array of in phase sources of period d_T, self-images at the Talbot distance $Z_T = 2d_T^2/\lambda_m$ where λ_m is the wavelength in the diffractive medium. It was found that at half the Talbot distance, the out-of-phase mode self images, while the in-phase mode image is displaced laterally by half of the period d_T [24] (see figure 7). Hence, if at half the Talbot distance an appropriate array of apertures allows the in-phase mode transmission, the out-of phase mode is effectively filtered [25].

The method was used by Mawst *et al* in 1989 [26], by fabricating an structure with the Talbot cavity integrated and the laser array monolithically integrated within a single device. The first results reported 70 mW/facet emitted power with a DL fundamental operation, stable up to twice the threshold current (~350 A).

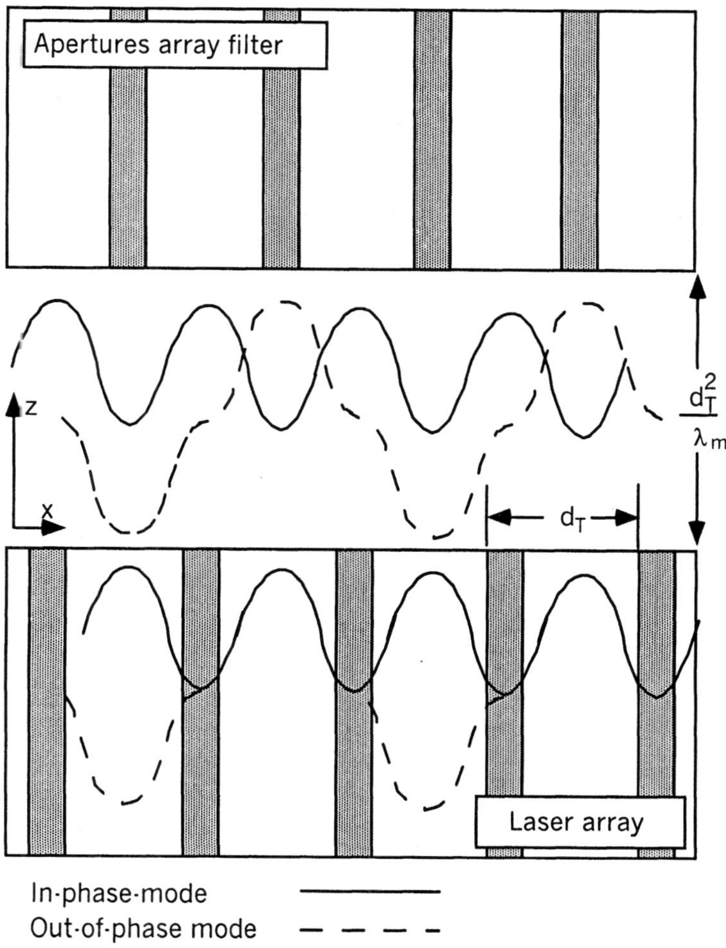

Figure 7: Top view schematic representation of a multichannel laser coupled array and spatial filter based on the Talbot effect. At half the Talbot distance, a periodic array of apertures displaced laterally by half of the period d_T, is set. That filters the out-of-phase mode. The wavelength of light in the diffractive medium is λ_m.

Late, novel structures of anti-guided (leaky-mode) lasers also emerged. These included improved one-dimensional (1-D) edge-emitting arrays and two dimensional (2-D) surface-emitting [27]. With the 1-D type arrays up to 1W of CW were delivered, in a 1.7X DL beam, with 70% of the energy in the main lobe. With the 2-D type array, up to 6W of peak pulsed power with 33% of fringes visibility were reported, from a 9-unit display array.

These authors predicted by that time a 16-unit 2-D arrays capable of locking in single spatial mode to output powers of more than 10W. Similar experiments were reported in progress to show a large 2-D array of 1500 elements capable of delivering up 50W of CW power, by placing the laser array in an external Talbot cavity [28]. These systems became too complex as they incorporated more and more elements.

2.2 Unstable Resonators

Simultaneously with the array systems development, other research areas were also of the interest of different working groups. Because of their capacity to deal with large lasing volumes and to eliminate high order lateral modes, unstable resonators (URs), were another important approach for obtaining high power coherent lasers. Optical resonators are divides in two types: stable and unstable. These concepts are illustrated in figure 8. A stable resonator (SR) is an optical cavity that confines light in a region of limited lateral width. On the contrary, an unstable resonator (UR) is an optical cavity for which light rays "walk" out of it. In the schematic diagrams of Fig. 8, the cavity type is determined by the configuration and shape of the mirrors. However, it shall be seen that other forms of unstable resonator cavity are possible.

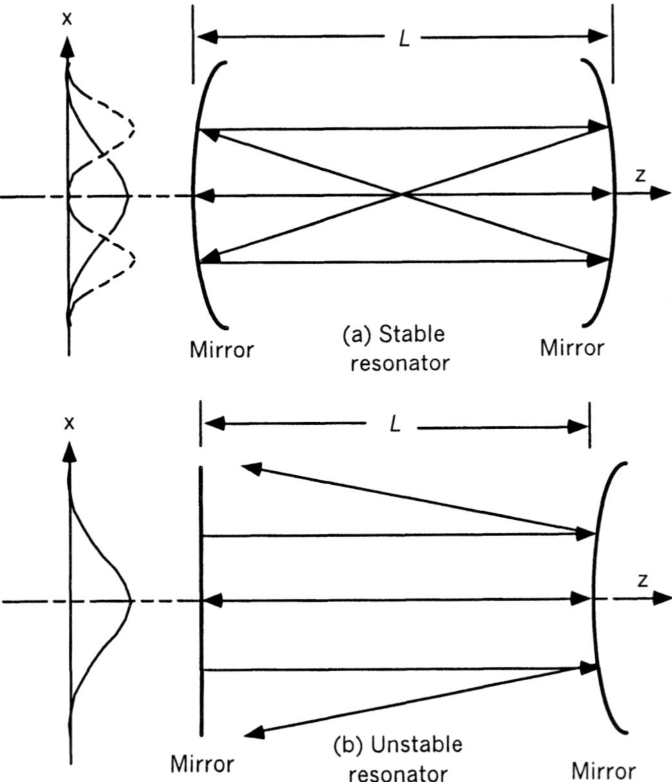

Figure 8: Schematic representation of the stable and unstable resonator concept. In (a) light bouncing back and forth inside the optical cavity can be trapped between the two mirrors. In (b), light which does not coincide with the center of the structure is deviated laterally. In this figure the ray along the z-axis represent the zero order mode. Rays off from the z-axis represent other order modes.

In the stable case, light bouncing back and forth inside the cavity is trapped, and it does not matter in which direction the light is moving (actually the direction itself is fixed by the radius of curvature of the mirrors). Consequently, the only way in which the light can escape is through one of the mirrors. Thus any ray which is not exactly parallel to the z-axis and travelling down the center line of the resonator will eventually bounce out of the resonator. Alternatively, the diverging element can be imagined to introduce lateral magnification or anti-guiding to the propagating mode.

The UR laser cavity provides less amplification and increases the losses for the high order modes because they are spatially more extended and leave the cavity after a few passes. In contrast the fundamental (on-axis) stays in the cavity for many more passes and is therefore amplified more. In this way discrimination in favor of the fundamental spatial mode, even for a large volume cavity, can be obtained.

The advantages of using URs in lasers for efficient low-order mode extraction has been an important study area in laser technology development for high power applications for many years [30]. A treatise on stable resonators (SRs) and URs, can be found in Chapters 19 and 22 respectively in Siegman's book on Lasers, [30] or the reader is also referred to the classical paper on laser resonators by Kogelnik and Li [31].

2.2.1 On Chip Unstable Resonators

First the UR approach started being used in gas and solid state lasers only, however, late it was shown that they were also suited to semiconductor lasers. Applying the UR principle to semiconductor lasers is however difficult because the small size of these devices. Nevertheless, using advanced semiconductor processing techniques it is possible to integrate unstable resonant cavities into diode lasers.

Several UR cavity configurations are possible in a semiconductor laser. One approach consists of etching concave profiles on one of the end facets of the laser chips either mechanically or by chemical means, to reproduce the cavity shown in figure 8b. In 1980, Bogatov *et al*, built an asymmetrical UR laser diode by grinding and polishing a cleaved facet into the shape of a curved surface, using diamond paste and a nylon string [32]. Figure 9 illustrates this type of UR. This early work demonstrated the effectiveness of the UR method by showing stable far field intensity distributions for a range of driven current from 3.9 A to 6.8 A. Even though the power level and the degree of spatial coherence reached were relatively low, the work encouraged the interest of many other researchers.

In 1985 Craig *et al*, and Salzman *et al*, reported fabrication of similar UR diode lasers using wet chemical etching and reactive-ion-etching (RIE) techniques, respectively [33, 34]. In the first case, devices with an etched mirror in one of the facets showed a threshold current of 1.0 A, independent of the mirror curvature and power levels up to 200 mW were reported. However specific values for spatial characteristics were not mentioned. In the second case curved mirrors in both facets of the laser allowed a higher loss cavity. A threshold current of 700 mA and external quantum efficiency of 22 % were reported, together with output powers of 350 mW. A "high degree of coherence" was achieved but was not quantified. Those works however encouraged even more the interest of working groups by showing the promise of obtaining high power with stable lateral mode characteristics simultaneously.

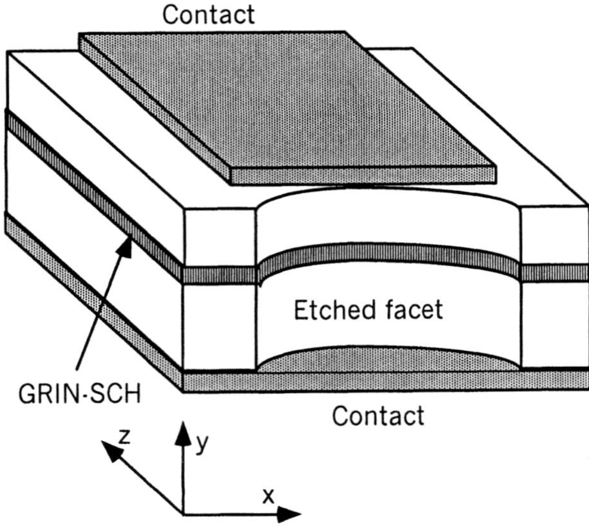

Figure 9: Schematic representation of the asymmetrical unstable resonator semiconductor laser. The section shown corresponds to the back facet where a concave mirror was produced by either mechanical abrasion, chemical etching or ion milling.

Later works by Larsgent *et al* [35], Tilton *et al* [36] and DeFreez *et al* [37] have shown the soundness of this approach by using improved techniques allowing more controllable parameters. The first group used RIE to anisotropically etch a curved facet at one end of the laser cavity, which was passivated with Si_3N_4. With this asymmetrical UR laser, they reported a threshold current of 250 mA, a slope efficiency of 0.25 W/A per facet and output powers up to 700 mW with a far field lobe at 2X DL. The second and third groups reported fabrication of a similar UR laser but using instead an even more controllable focused-ion-beam micromachining (FIBM) technique to implement the curved facet mirror. In the first case total > 1W at nearly DL regime were reported; in the second, powers as high as 6 W for pulsed operation and 1.2 W in CW were obtained.

2.2.2 Advanced Unstable Resonator Designs

A different approach that was believed could lead to a more manufacturable device, was to integrate the diverging elements inside the cavity by using a two part waveguide. The first waveguide was an asymmetric GRIN-SCH with the quantum well gain region. The lasing mode also couples to a secondary waveguide located beneath or above the primary waveguide, in which diverging elements are incorporated to provide the unstable resonator action. Figure 10 illustrates this idea.

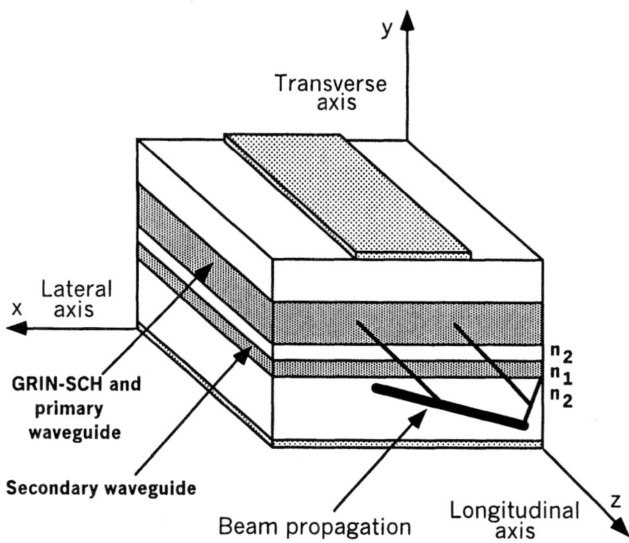

Figure 10: Illustration of the two part waveguide laser concept. Laser light is originated at the primary waveguide. Lasing modes couple towards the secondary waveguide where diverging elements are incorporated that transform the medium into a unstable resonator.

One example of this was the regrown lens train (RTL) laser, implemented by S. T. Srinivasan *et al* at the CHTM of UNM [38], that incorporates an unstable resonator in the secondary waveguide by etching negative diverging lens elements into a GaAs layer, above the GRINS-SCH, and a subsequent regrowth of the lower index p-type cladding layer and contact layers. See figure 11. The negative or diverging lenses magnify the modes so that, eventually these of higher orders are deviated out of the gain region more rapidly than the lower ones, thus causing the desired modal discrimination. Or in other words, the medium becomes an unstable resonator. With this structure, a threshold current density of 400 A/cm^2 and a power of 490 mW/facet, with a far field lobe of 1.2X DL, was achieved.

The shaped unstable resonator (SHUR) laser, also uses a two part waveguide and is therefore in some respects, similar to the RLT laser (figure12). In the case of the SHUR laser however, the diverging element is a non-planar guide that creates a parabolic lateral index variation and is incorporated into the secondary waveguide below the main GRIN-SCH region. This secondary waveguide is a GaAs layer that after being photoetched is filled in with a lower index material ($Al_xGa_{1-x}As$) [39]. The effective refractive index of the medium varies almost linearly with the thickness of the GaAs layer, being higher towards the sides of the device, as shown in the lower part of Fig. 12. This forces the light which is traveling parallel to the z direction but out of axis, to be deviated out of the structure and so the optical cavity becomes an unstable resonator. With this design the following data were achieved: up to 770 mW /facet (47 % in the main lobe), pulsed, with an external quantum efficiency of 66 %. The far field showed a lobe 2X the DL at FWHM, while the threshold current was 1.4 A [40], due to the large contact area used.

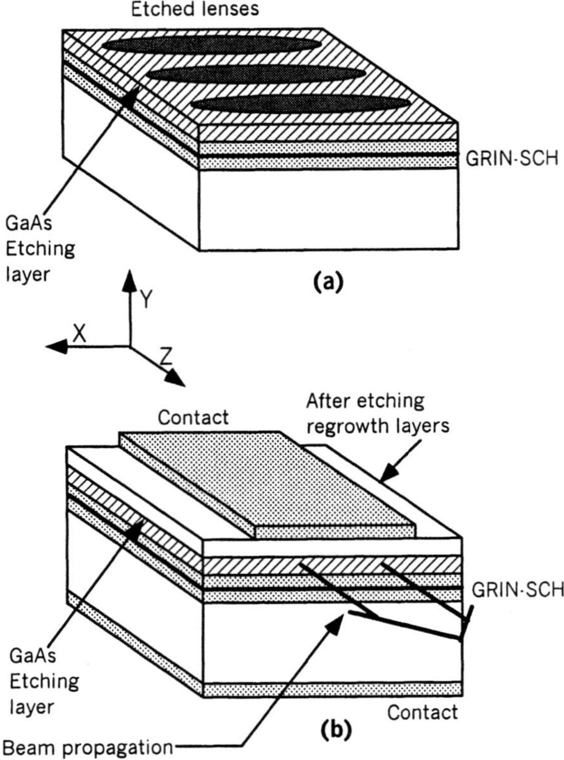

Figure 11: In the regrown lens train (RLT) laser, shown schematically, an unstable resonator is created by etching in a GaAs layer above the GRIN-SCH. In (a) many lens shaped openings are etched that later will be infilled with $Al_xGa_{1-x}As$. In (b), after regrowth, metallic contacts are deposited on top and bottom of the whole structure.

2.3 The Master Oscillator Power Amplifier (MOPA) approach

By the middle 90's, emerged one of the more elegants concepts to achieve high power from a semiconductor laser, based on the so-called: master oscillator/power amplifier (MOPA) configuration [41,42,43].

The MOPA consists of master of oscillator working at relatively low power, coupled to an amplifier that rises the power level by a factor from 10 to 100, while maintaining the high spatial and spectral characteristics of the original signal (see figure 13a). A variation of this basic configuration is the flared MOPA (figure 13b) in which the master oscillator can be a low power (~50 mW) and the power amplifier is tapered with a narrow (~10 μm) input and flared output (> 200 μm). So far this device has demonstrated the highest single-spatial-mode CW power (> 3W) obtained with a diffraction limited far field distribution. In a similar device using a broad-area amplifiers injected with ever higher power, huge powers were mentioned by the same authors (~22 W pulsed and 11.5 W CW), however no mention of the degree of coherence obtained was done.

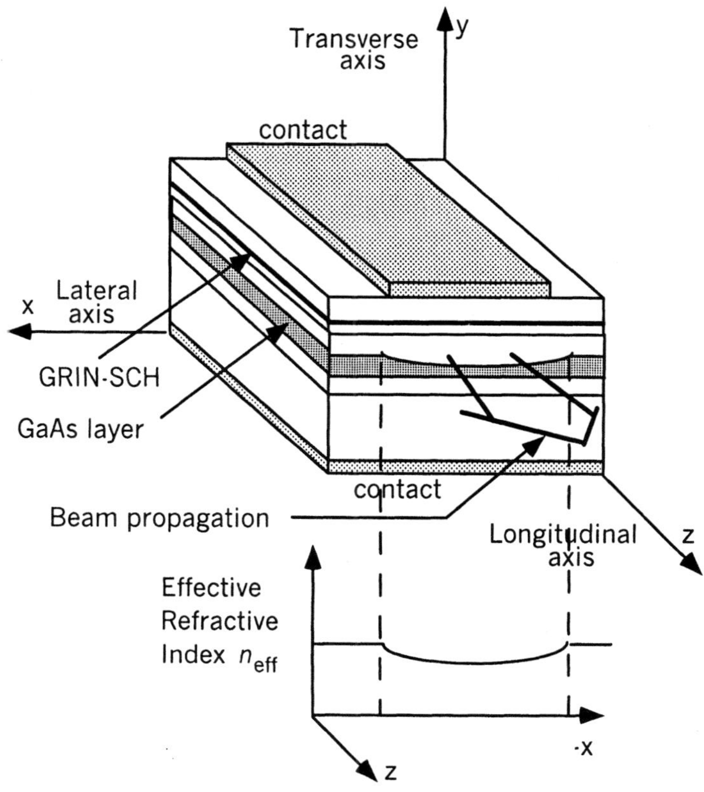

Figure 12: Schematic representation illustrating the shaped unstable laser resonator (SHUR) laser concept. The curved area represents the photoetched GaAs layer (secondary waveguide) on top of which an $Al_xGa_{1-x}As$, is regrown and then the GRIN-SCH (primary waveguide) is completed. The medium becomes an unstable resonator after this operation due to the negative lateral effective index variation.

2.4 Conclusions and Comments

In the paragraphs above, an overview as complete as possible has been presented, on the state-of-the-art of the existing methods, structures and designs, to achieve high coherent power from a laser diode, since its invention to the present. The huge amount of published works in the area, makes almost impossible a full compilation of the subject and the speed at which the discovering of new materials and structures is performed, makes more difficult the task. However is for this reason, and for the diversity of designs and developed works but essentially for the applications, that we can see the interest of this subject among the researchers groups and is for this that we can assure, for the time being this interest has not come to an end.

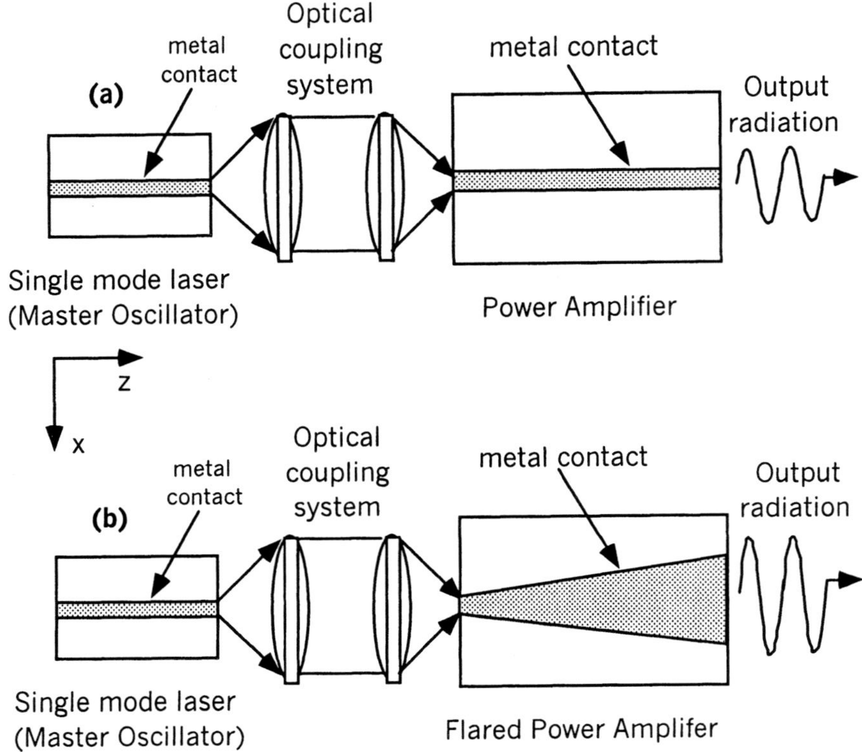

Figure 13: Top view illustration of the master oscillator power amplification (MOPA) concept, one of the most effective methods to achieve high power and coherent radiation from a semiconductor laser. In (a), a low power coherent laser (Master Oscillator) is coupled to a gain medium (Power Amplifier) that amplifies the signal. In (b), a tapered gain region is used, that maximizes power extraction efficiency.

The systems here mentioned work basically in the infrared. Systems for different wavelengths are not included since the methods to obtain high power were developed for structures based in GaAs as emitter and before designs for other wavelengths were proposed as structures based on GaN [44,45] of very recent proposal. Is it of course possible that the methods here described, could be applied to systems in a different range of wavelengths.

Recent publications, show in a period of 2 to 3 years to date, that nowadays, the effort of the working groups is more well focused to the improvement of the existing systems and to their applications for example in areas such as: sensors, spectroscopy, optical processing and virtual games, etc., [46,47,48], using very sophisticated structures and for another wavelength ranges, other than the infrared, but that essentially no different methods than those here mentioned, has been proposed for achieving high coherent power. The matter of the applications has not been examined in this work, but could be the subject of a future work.

References

[1] R. N. Hall, G. E. Fenner, J. D. Kingsley, T. J. Soltys and R. O. Carlson, *Phys. Rev. Lett.* 9, 366 (1962).

[2] M. I. Nathan, W. P. Dumke, G. Burns, F. H. Dill, Jr. and G. Lasher, *Appl. Phys. Lett.* 1, 62 (1962).

[3] T. M. Quist, R. H. Rediker, R. J. Keyes, W. E. Krag, B. Lax, A. L. McWhorter and H. J. Zeigler, *Appl. Phys. Lett.* 1, 91 (1962).

[4] N. Holonyak Jr. and S. F. Bevacqua, *Appl. Phys. Lett.* 1, 82 (1962).

[5] L. Figueroa Ed., Proc. of SPIE, Vol.893, (1988).

[6] H. C. Casey Jr. and M. B. Panish, *Heterostructure Lasers* (Part A), Chapter 2, Academic Press, New York, (1978).

[7] A. Siegman, *Lasers*, Chapter 14, University Science Books, Mill Valley CA, (1984).

[8] I. Hayashi, M. B. Panish, P. W. Foy and S. Sumski, *Appl. Phys. Lett.* 17, 109 (1970).

[9] R. D. Dupuis and P. D. Dapkus, *Appl. Phys. Lett.* 32, 473 (1978).

[10] R. D. Dupuis, P. D. Dapkus, N. Holonyak Jr. and R. M. Kolbas, *Appl. Phys. Lett.* 35, 487 (1979).

[11] W. T. Tsang, *Appl. Phys. Lett.* 39, 134 (1981).

[12] S. Hersee, M. Baldy, P. Assenat, B. De Cremoux and J. P. Duchemin, *Electron Lett.* 18, 618 (1982).

[13] D. R. Schifres, R. D. Burnham and W. Streifer, *Appl. Phys. Lett.* 33, 1015 (1978).

[14] D. R. Scifres, W. Streifer and R. D. Burnham, *Appl. Phys. Lett.* 34, 259 (1979).

[15] D. R. Scifres, R. D. Burnham and W. Streifer, *Appl. Phys. Lett.* 41, 118 (1982).

[16] D. R. Scifres, R. D. Burnham and W. Streifer, *Appl. Phys. Lett.* 41, 1030 (1982).

[17] D. E. Ackley and R. W. H. Engelmann, *Appl. Phys. Lett.* 39, 27 (1981).

[18] D. E. Hackley, *IEEE J. Quantum Electron.* QE-18, 1910 (1982)

[19] D. Botez, L. J. Mawst, P. Hayashida, G. Peterson and T. J. Roth, *Appl. Phys. Lett.* 53, 464 (1988).

[20] L. J. Mawst, D. Botez, M. Jansen, T. J. Roth, C. Zmudzinski, C. Tu and J. Yun, *Proc. of SPIE*, Vol.1634, p.2 (1992).

[21] D. Botez, *IEE Proc.-J*, Vol.139, p.14 (1992).

[22] T. S. Shiau, S. Sun, C. F. Schaus, K. Zheng and G. R. Hadley, *IEEE Phot. Tech. Lett.* 2, 534 (1990).

[23] D. Mehuys, J. Major Jr. and D. F. Welch, *Proc. of SPIE*, Vol.1850, p.2 (1993).

[24] A. A. Golubentsev, V. V. Likhanskii and A. P. Napartovich, *Sov. Phys. JETP* 66, 676 (1987).

[25] C. Roychoudhuri, E. Siebert, F. D'amato, S. Macomber and R. Noil, *LEOS '88 Annual Meeting Proc.* Paper FE.4, p.447 (1988).

[26] L. J. Mawst, D. Botez, T. J. Roth, W. W. Simmons, G. Peterson, M. Jansen, J. Z. Wilcox and J. J. Yang, *Electron. Lett.* 25, 365 (1989).

[27] D. Botez, M. Jansen, L. J. Mawst, T. J. Roth, S. S. Ou, J. J. Yang, C. Zmudzinski, P. Hayashida and M. Seargant, *Diode Laser Technology Program (DLTP) Proc.*, Sixth Annual Conference, Albuquerque NM, (1993).

[28] R. Waarts, D. Nam, S. Sanders, D. Welch and S. Scifres, *Diode Laser Technology Program (DLTP) Proc.*, Sixth Annual Conference, Albuquerque NM, (1993)

[29] A. E. Siegman, *Appl. Opt.* 13, 353 (1974).

[30] A. E. Siegman, *Lasers*, University Science Books, Mill Valley CA, (1986).

[31] H. Kogelnik and T. Li, *Proc. of the IEEE*, Vol.54, p.1312 (1966).

[32] A. P. Bogatov, P. G. Elisev, M. A. Man'ko, G. T. Mikaelyan and Y. M. Popov, *Sov. J. Quantum Electron.* 10, 620 (1980).

[33] R. R. Craig, L. W. Casperson, O. M. Stafsudd, J. J. J. Yang, G. A. Evans and R. A. Davidheiser, *Electron. Lett.* 21, 62 (1985).

[34] J. Salzman, T. Venkatesan, R. Lang, M. Mittelstein and A. Yariv, *Appl. Phys. Lett.* 46, 218 (1985).

[35] C. Largent, D. Gallant, J.Yang, M. Allen and M. Jansen, *Proc. of SPIE*, Vol.1418, p.40 (1991).

[36] M. L. Tilton, G. C. Dente, A. H. Paxton, J. Cser, R. K. DeFreez, C. E. Moeller and D. Depatie, *IEEE J. Quantum Electron.* 27, 2098 (1991).

[37] R. K. DeFreez, Z. Bao, P. D. Carleson, M. K. Felisky and C. Largent, *Proc. of the SPIE*, Vol. 1850, p.75 (1993).

[38] S. T. Srinivasan, C. F. Schaus, S. Z. Sun, E. A. Armour, S. D. Hersee, J. G. McInerney, A. H. Paxton and D. J. Gallant, *Appl. Phys. Lett.* 61, 1272 (1992).

[39] S. Guel Sandoval, *High-Power Coherent Semiconductor Laser with Continuous Photoetched Shaped Unstable Resonator, Ph. D. Dissertation,* University of New Mexico, Albuquerque, NM (1994).

[40] S. Guel Sandoval, A. H. Paxton, S. T. Srinivasan, S. Z. Sun, S. D. Hersee, M. S. Allen, C. E. Moeller, D. J. Gallant, G. C. Denty and J. G. McInerney, *Appl. Phys. Lett.* 66, 2048 (1995).

[41] D. Mehuys, D. F. Welch and L. Goldberg, *Electron. Lett.* 28, 1944 (1992).

[42] R. Parke, D. F. Welch, A. Hardy, R. Lang, D. Mehuys, S. O'Brien, K. Dzurko and D. Scifres, *IEEE Phot. Tech. Lett.* 5, 297 (1993).

[43] R. J. Lang, S. O'Brien, D. Mehuys and D. F. Welch, *Diode Laser Technology Program (DLTP) Proc.*, Sixth Annual Conference, Albuquerque NM, (1993).

[44] S. Nakamura, *IEEE J. of Selected Topics in Quantum Elect.* 3, 435 (1997).

[45] S. P. Denbaars, *Proc. of the IEEE*, Vol. 85 No. 11, p.1740, (1998).

[46] J. N. Walpole, *Opt. and Quantum Electron.* 28, 623 (1996).

[47] K. J. Linden Ed., *Proc of SPIE*, Vol. 3000, (1997).

[48] M. Fallahi and S.C. Wang Eds., *Proc. of SPIE*, Vol. 3004, (1997).

[49] G. P. Agrawal and N. K. Dutta, *Long-Wavelength Semiconductor Lasers*, Chapter 2, Van Nostrand Reinhold, New York, (1986).

[50] B. E. A. Saleh and M. C. Teich, *Fundamentals of Photonics*, Chapter 16, Wiley & Sons, New York, (1991).

[51] H. C. Casey Jr. and M. B. Panish, *Heterostructure Lasers,* (Part A), Chapter 2, Academic Press, New York, (1978).

[52] G. O. Reynolds, J. B. DeVelis, G. B. Parrent Jr. and B. J. Thompson, *Physical Optics Notebook: Tutorials in Fourier Optics,* Chapters 10 and 11, SPIE Optical Engineering Press, Bellingham, WA, (1989).

Appendix A

This appendix describes the techniques commonly used to characterize the spatial coherence and power in a semiconductor laser.

A. Light-Current characteristics

One of the primary characteristics of interest in a semiconductor laser is the light power emitted by one facet as a function of the injected [49]. The resulting graph is commonly known as the P-I curve (see figure 14). Important information can be drawn from this data as for example the threshold current I_{th}, value which marks the transition from non-lasing to lasing emission. Another basic quantity, invariably used to characterize the performance of a semiconductor laser is the quantum efficiency (internal η_i, and/or external η_e), that indicates how efficient the laser is in converting electrical energy into optical energy.

The internal quantum efficiency η_i in an injection semiconductor laser or a light emitting diode (LED), is defined as the fraction of the injected electron flux (this is: *i/e*, where i is the injected current and e is the electron charge, in units: *electrons/second*) that is converted into photon flux internally Φ. In other words, η_i is the ratio of the number of electron-hole pairs that recombine radiativelly and generate photons, inside the diode, to the number of electrons injected into the diode [50].

The internal photon flux Φ, is expressed as

$$\Phi = \eta_i \frac{i}{e} \tag{A.1}$$

The output photon Φ_o is related to the internal photon flux Φ, by:

$$\Phi_0 = \eta_e \Phi = \eta_e \eta_i \frac{i}{e} \tag{A.2}$$

where η_e is the overall transmission efficiency with which the internal photons can be extracted from the structure. If the external differential quantum efficiency η_d is defined as

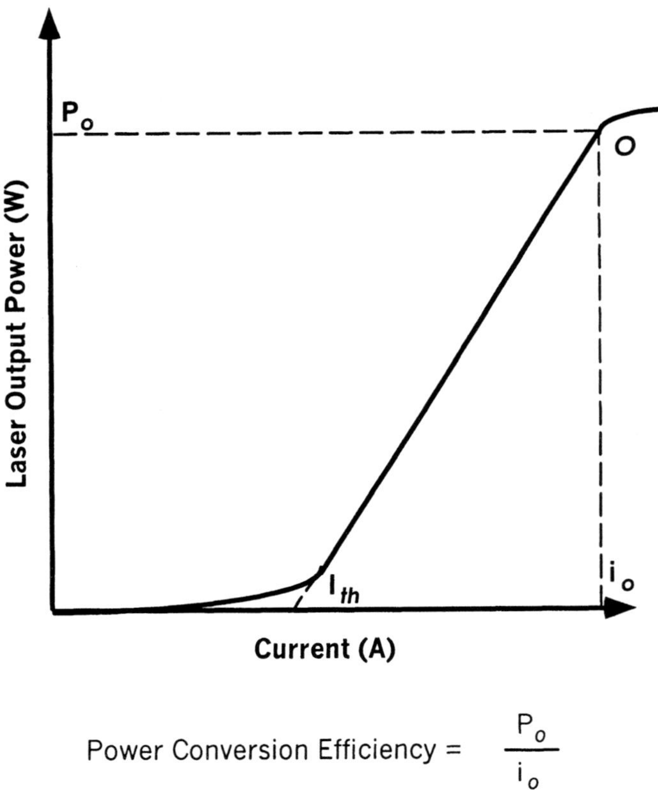

$$\text{Power Conversion Efficiency} = \frac{P_o}{i_o}$$

Figure 14: Ideal characteristic *P-I* curve for a laser diode. The slope of the curve after threshold, in the linear region, multiplied by a constant, is approximately equal to the external quantum efficiency and so is the power conversion efficiency defined as the optical output power to the applied electrical current.

$$\eta_P = \eta_P \eta_t \tag{A.3}$$

then, the output photon flux can be written as

$$\Phi_o = \eta_P \frac{i}{e} \tag{A.4}$$

Since each photon has an energy $h\nu$, the laser diode output optical power P_o is related to the output photon flux by

$$P_o = h\Phi_o = \eta_P h \frac{i}{e} \tag{A.5}$$

The external quantum efficiency .. can be measured by the slope of the *P-I* curve before saturation is reached (Fig. 14). This slope is also called the laser responsivity *R* and is equal to the ratio of the emitted optical power to the injected current with units *Watts/Amp*. From Fig. 14, we obtain:

$$R = \frac{P_o}{i_o} = \frac{h\nu_o}{i_o} \tag{A.6}$$

Using Eq. (A.4), R can be expressed as

$$R = \eta_d \frac{h\nu}{e} \tag{A.7}$$

For $h\nu \approx eV$, where eV is the energy gap, R becomes

$$R = \eta_d \frac{1.24}{\lambda} \tag{A.8}$$

Hence the responsivity R of a laser (*Watts/Amp.*), can be expressed as η_d times a factor that depends on the emission wavelength λ and vice-versa.

Another measure of performance is the overall quantum efficiency η_p, also called the power-conversion efficiency, defined as the ratio of the obtained optical power P_o to the applied electrical power.

$$\eta_p = \frac{P_o}{i_o V} \tag{A.9}$$

Substituting (A.5), results

$$\eta_p = \eta_d \frac{h\nu}{eV} \tag{A.10}$$

Again taking $h\nu \approx eV$, then

$$\eta_p \approx \eta_d \tag{A.11}$$

This is, the power conversion efficiency and the external quantum efficiency are approximately equal.

Figure 15 shows schematically the experimental set-up used for *P-I* curve characterization.

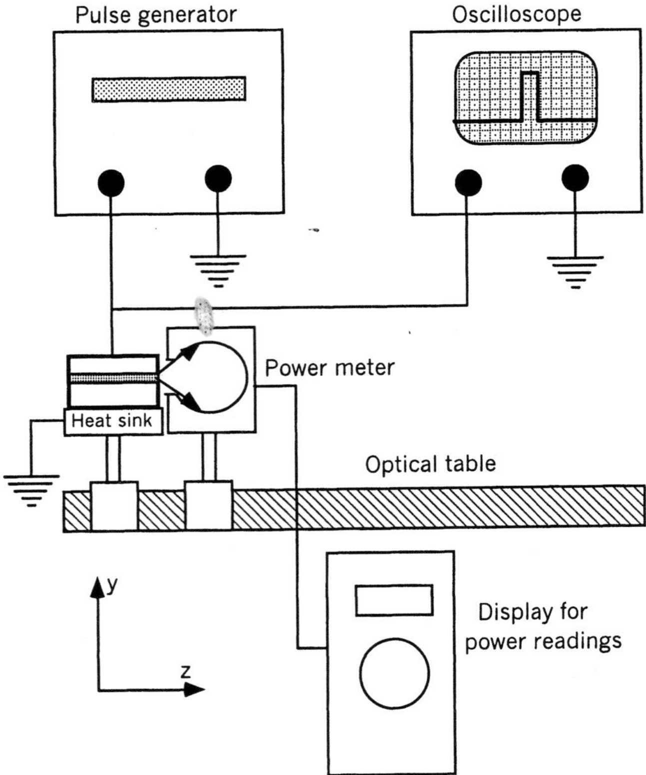

Figure 15: Schematic representation of the used for *P-I* curve measurement. A power meter is set directly in front of the laser, while current is made to vary. Often pulsed lasers are measured, so an oscilloscope is necessary to characterize pulsed current.

B   Spatial-Mode characteristics.

A semiconductor laser emits light in the form of a narrow elliptical cross section spot. See figure 16. The spatial intensity distribution of the emitted light near to the laser is called the near field. While the light beam intensity distributions far enough from the laser that the laser emitting area can be neglected, is known as far field. Due to purely diffraction effects, these two fields formed by the superposition of several transversal and lateral modes, are different.

The final dimensions of the elliptical spot area, and values of divergence angles $\theta_\perp$ and θ_Π, both perpendicular and parallel to the junction plane respectively, are also important parameters related to the spatial modal characteristics of the laser [51]. For example to accomplish an adequate coupling between the beam and an optical fiber, is necessary to know in detail such values.

The laser can be imaged to have a light virtual source of width s (as shown in figure 17), that emits diverging light creating a near field of width S. The condition for the field amplitude to have a minimum at any side of S, is the optical path difference to be:

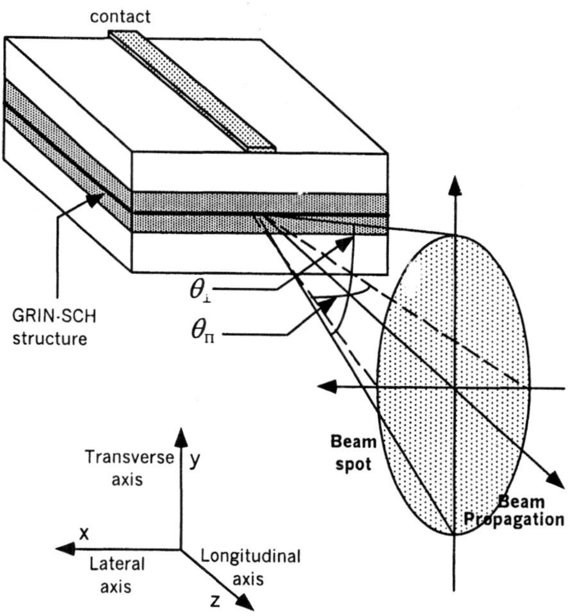

Figure 16: Representation of the far-field emission from a stripe geometry GRIN-SCH laser. In this schematic figure $\theta_\perp$ and θ_Π are the full angles in the directions perpendicular to and along the junction plane.

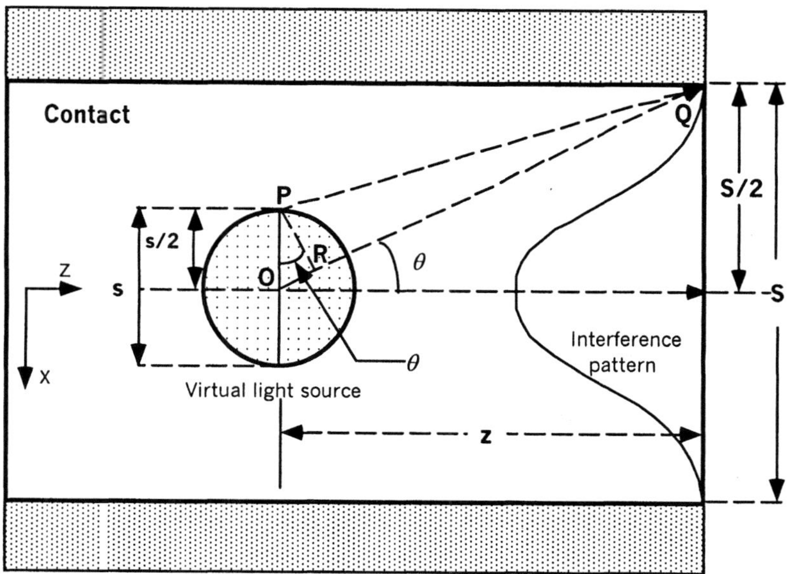

Figure 17: Ideal virtual light source inside a laser diode as generated by the out coming coherent beam. To determine its size s, one assumes the ideal case, called diffraction limit, in which the first minimum of the interference pattern between waves generated at *P* and *O*, coincide with the size of the emitting aperture.

$$\tilde{O}\tilde{Q} - \tilde{P}\tilde{Q} = \frac{\lambda}{2} \tag{A.12}$$

where λ is the wavelength of the emitted light. Besides

$$\tilde{O}\tilde{Q} - \tilde{P}\tilde{Q} = \frac{s}{2} sen\,\theta = \frac{\lambda}{2} \tag{A.13}$$

On the other hand, for small θ:

$$\frac{S}{2} \bullet z\,sen\,\theta \tag{A.14}$$

where z is the position of the light source inside the laser and S is the. Combining Eqs. (A.13) and (A.14), results:

$$s = \frac{2\lambda z}{S} \tag{A.15}$$

called the diffraction limit (DL) ideal virtual light source size, inside the diode. If the diverging output is focused, then the width w of the image is the magnified image of this virtual source s. So for the DL ideal case:

$$w = ms \tag{A.16}$$

where m stands for the imaging lens magnification, (see figure 18).

Commonly the width of the main lobe measured at the point where the intensity decreases to the half of the maximum value, is taken as reference. This point was already described as the full width at the half maximum (FWHM) value. If the width of the source s is measured at the FWHM point, this values is found to be:

$$s^{\odot} = 0.866\frac{\lambda z}{S} \tag{A.17}$$

assuming s behaves in intensity as an ideal *sinc²* function. Then this is defined as the ideal DL value of the source s, (limited only by diffraction).

Similarly at the far field distribution, the FWHM of main lobe, ideally should be equal to the FWHM of the virtual source times the magnification m of the imaging system, this is:

$$w^{\odot} = ms^{\odot} \tag{A.18}$$

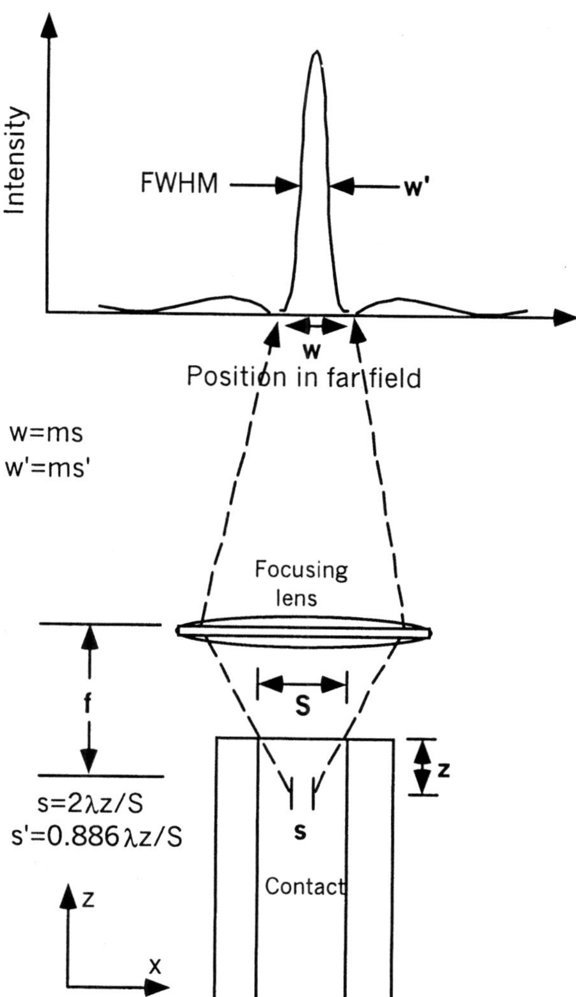

Figure 18: Illustration of the relation between the virtual light s inside laser and the width w of the far field lobe. In the ideal case, the width w' where the lobe intensity drops approximately to the half of its maximum value, must be equal to s' times the magnification m of the imaging system; where s' is the FWHM of the virtual source. S is the size of the emitting layer fraction, z is the position of s inside the diode and f is the lens focal length.

If the virtual source or near field is not perfectly coherent, this will increase the width of the far field image. So the FWHM of the far field main lobe, is a good measure of the laser spatial coherence. Typically at FWHM $\theta_\perp$ ranges from 30° to 65°, whereas θ_Π is in the order of 10°, where $\theta_\perp$ and θ_Π were defined in Fig. 16.

Figure 19, shows an scheme of the optical system used for far field characterization. A lens focuses the light coming from the laser onto a CCD camera. The CCD camera is connected to a TV monitor that displays an image of the near field (emitting facet) or the far field (virtual source), distributions and to a computer that analyzes this information.

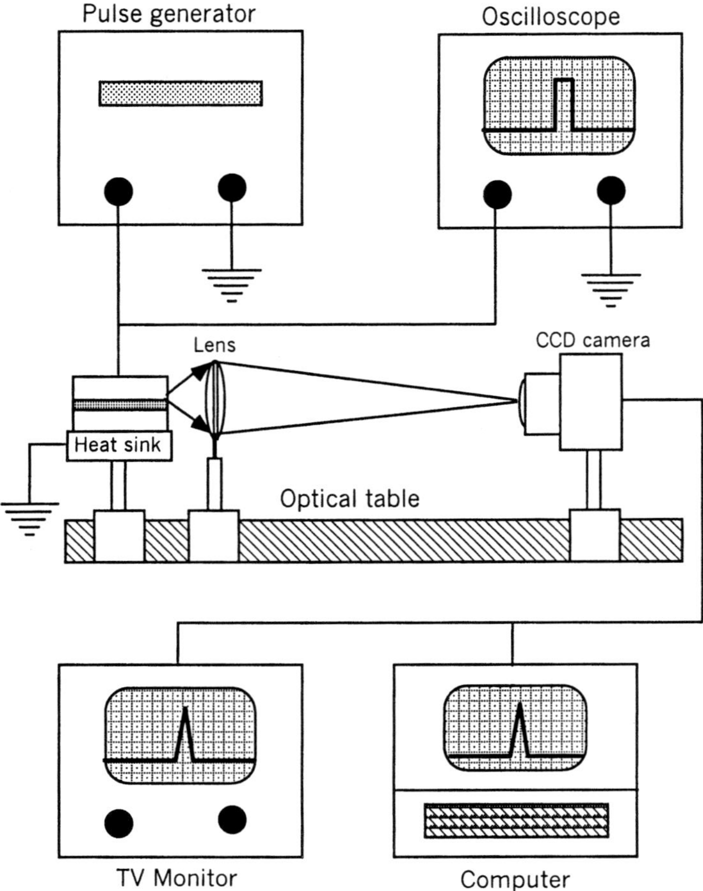

Figure 19:   Experimental set-up used for far field characterization measurements.
Emitted radiation from laser is focused onto a CCD camera connected to a TV monitor
and to a computer to be analyzed.

Another way of measuring spatial coherence of the laser output is by imaging the
near field on a double slit and measuring the visibility of the interference pattern as
shown in figure 20. The visibility of an interference fringe pattern is defined as:

$$V = \frac{I_{Max} - I_{Min}}{I_{Max} + I_{Min}}$$

(A.19)

where I_{Max} and I_{Min} are the maximum and minimum intensities in the fringe pattern.

The visibility of an interference pattern, thus defined, can be demonstrated to be
equal to the modulus of the complex degree of coherence between the wavefronts
producing such pattern [52]. If those wavefronts have a common source, the visibility of
the fringes is a direct measure of the degree of spatial coherence of that source.
Consequently it can be used to evaluate the quality of the laser beam: the better the
spatial coherence of the laser the narrower the lobe in the far field and the higher the

visibility of the fringes. However, caution should be used when applying this procedure, since a damaged facet can give poor visibility fringes even if the laser has good coherence.

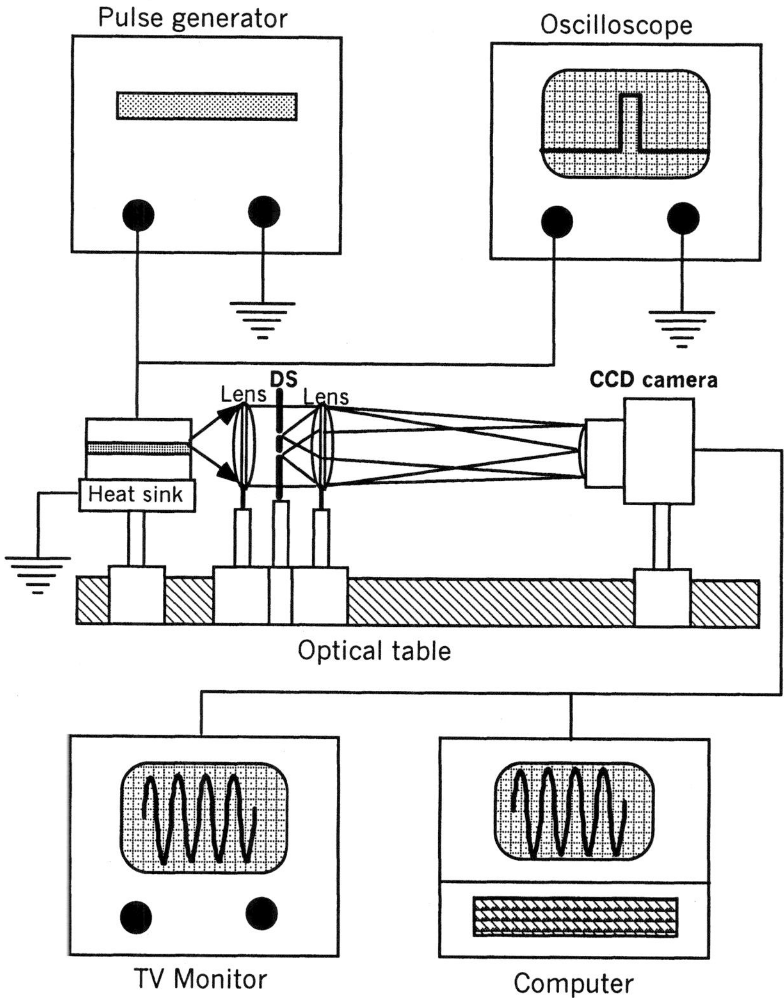

Figure 20: Schematic visibility measurement set-up. Interference Young fringes are projected onto a CCD camera connected to a TV monitor. The visibility of the fringes is a direct measure of the spatial coherence degree in the laser beam.

Carrier Transport Properties of (n)nc-Si:H/(p)c-Si Heterojunction

Y. C. Peng[1,2], G. S. Fu[3], G. Y. Xu[4], Y. L. He[5], M. Liu[6], and Y. X. Li[2]

1). College of Electronic and Informational Engineering, Hebei University, Baoding 071002,China

2). State Key Laboratory of superlattice and Microstructures, Institute of Semiconductors, Chinese Academy of Sciences, Beijing 100083, China

3). College of Physical Science and Technology, Hebei University, Baoding 071002, China

4). Shanghai Institute of Metallurgy, Chinese Academy of Sciences, Shanghai 200050, China

5). Department of Physics, Nanjing University, Nanjing 210093, China

6). Microelectronic Center, Chinese Academy of Sciences, Beijing 100029, China

Abstract

Phosphor-doped nanocrystalline silicon ((n)nc-Si:H) film was successfully grown on p-type oriented crystal silicon ((p)c-Si) substrate by conventional enhanced chemical vapor deposition. The film was obtained using high H_2 diluted SiH_4 as a reaction gas source and using PH_3 as the doping gas source of phosphor atoms. Furthermore, the heterojunction diodes were also fabricated by (n)nc-Si:H film and (p)c-Si substrate. I-V characteristics are investigated in the temperature range of 230-420K. The experimental results demonstrated that (n)nc-Si:H/(p)c-Si heterojunction is a typical abrupt heterojunction with good rectifying properties and temperature stability. Carrier transport mechanism is tunneling-recombination model at forward bias voltage. At low bias voltage (V_F < 0.8V), the current is determined by recombination at (n)nc-Si:H side of the space charge region, while at higher bias voltage (V_F >1.0V), the current is dominated by electron tunneling processes. The heterojunction diodes have high reverse breakdown voltage (>-75V) and low reverse current ($\approx$on the order of nA)

Keywords: (n)nc-Si:H/(p)c-Si heterojunction, energy band model, carrier transport mechanism, temperature stability

1 Introduction

Recently, electrical properties of various silicon-based heterojunctions are of increasing interest. Among them are heterojunctions formed by amorphous silicon(a-Si:H), microcrystalline silicon(μs-Si:H), porous silicon (ps-Si) and nanocrystalline silicon (nc-Si:H) together with crystal silicon (c-Si), etc.[1-4] This is due to the fact that they have potential applications in the new electronic devices. Different carrier transport mechanisms have been proposed by measuring current-voltage (I-V) characteristics of the heterojunction diodes. Marsal et al. [5] presented an analysis of carrier transport mechanism on a-Si:H/c-Si heterojunction, and confirmed that at low bias voltage (V<0.4V), the current is determined by recombination at the amorphous side of the space charge region, while at higher voltage(V>0.6V), the current becomes space charge limited. Narasimhan et al. [6] proposed the carrier transport mechanism of ps-Si/c-Si heterojunction. It has been found that the I-V characteristics of the heterojunction have weaker temperature dependence when resistivity of a c-Si substrate is larger than $4\,\Omega \cdot cm$. The experimental results were explained by means of electronic tunneling model.

We have fabricated phosphor-doped nanocrystalline silicon ((n)nc-Si:H) film by plasma chemical vapor deposition(PECVD). Furthermore, the heterojunction diodes were also fabricated by (n)nc-Si:H film and c-Si substrate. I-V characteristics are investigated in the temperature range of 230-420K. In this work, we present an analysis on the temperature dependence of the I-V characteristics. A detailed discussion of carrier transport mechanisms follows.

2 Experimental

(n)nc-Si:H/(p)c-Si heterojunnctions have been studied using a diode structure. The cross-sectional configuration of the sample is shown in Fig.1. The substrate is p-type silicon <100> oriented with an average resistivity of $1\,\Omega \cdot cm$. First, a 200nm thick SiO_2 layer was grown by conventional thermal oxidation at temperature of 1020 C. The SiO_2 layer was etched and patterned by photolithography to make an array of square holes of $100\,\mu m$. After appropriate treatment, a 150nm thick (n) nc-Si:H film was deposited on the array configuration by PECVD method. The film is obtained using high H_2 diluted SiH_4 as a reaction gas source and using PH_3 as the doping gas source of phosphor atoms. The nanocrystalline silicon grains size and doping concentrations in the film are about 3nm and $3\text{-}4\times10^{17} cm^{-3}$, respectively. Then, the samples were followed by thermal annealing treatment in the N_2 ambient air. The outer layer of (n)nc-Si:H film in the square holes was removed by etching and photolithography, leaving only (n)nc-Si:H layer in the hole bottoms. In the last step, Al electrodes of $120\,\mu m \times 120\,\mu m$ were evaporated on each square. The microstructure of the film was examined by high resolution transmission microscopy (HREM), to confirm that the film was really nanocrystalline in structure. The

I-V characteristics were measured using a H4140 at different temperatures within the range of 230-420K.

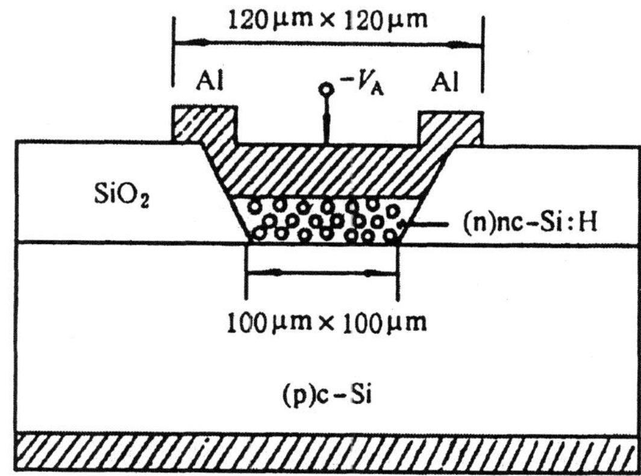

Figure 1: Cross-sectional view of a (n)nc-Si:H/(p)c-Si heterojunction diode

3　Results and Discussion

3.1 Energy Band Diagram of the Heterojunction

The energy band diagrams of an ideal (n)nc-Si:H/(p)c-Si heterojumction are shown in Fig.2. The each equilibrium energy band diagrams of (n)nc-Si:H film and (p)c-Si substrate are shown in Fig.2(a) when the heterojunction was not formed. Eg_1=1.12eV and E_{F1} are energy gap and Fermi level of (p)c-Si substrate, Eg_2 and E_{F2} are energy gap and Fermi level of (n)nc-Si:H film, respectively. It should be noted that the energy band diagram of (n)nc-Si:H film is a simplified schematics diagram. So far, since no reliable values of Eg_2 for (n)nc-Si:H film have been reported, it is replaced by optical energy gap Eg^{opt}. Our previous studies indicated that the values of Eg^{opt} are related to both hydrogen contents and volume fractions of nanocrystalline silicon in (n)nc-Si:H film, and are estimated at 1.70-1.85eV from ultraviolet spectra and the Taue curve[7-8]. These values are in good agreement with the calculated results for optical energy gap of silicon nanoclusters[9]. Fig.2(b) shows the thermal equilibrium energy band diagram of (n)nc-Si:H/(p)c-Si heterojunction. It is clearly seen that the conduction band edge discontinuity is generated in the interface region of heterojunction. At the same time, a sharp energy barrier ΔE_c is formed, due to the energy gap difference between (n) nc-Si:H film and (p)c-Si substrate. The energy band diagrams of the heterojunction under forward and reverse bias voltage are shown in Fig.2(c) and (d), respectively. According to above energy band model, we can give an analysis for the J-V characteristics of the heterojunction under different temperature ranges and bias conditions, and explain carrier transport mechanism.

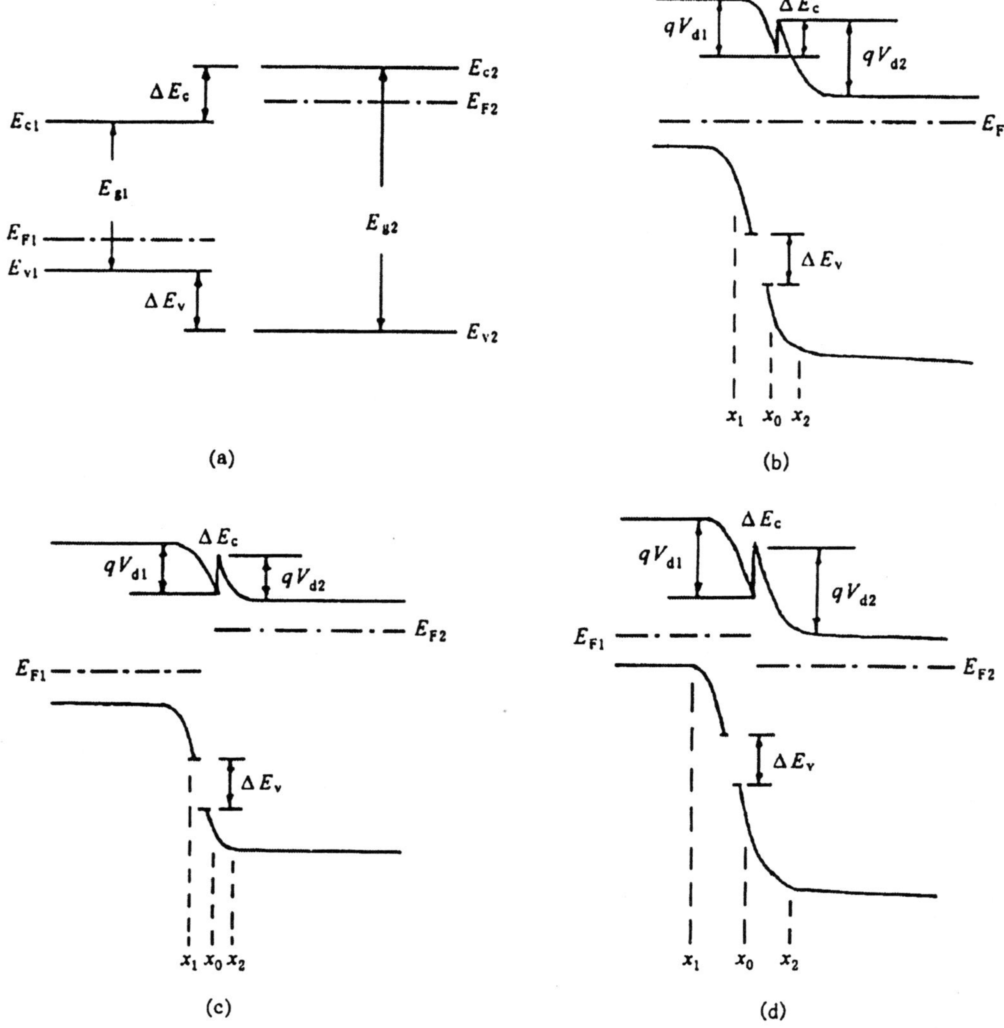

Figure 2: Energy band diagram of the (n)nc-Si:H/(p)c-Si heterojunction

3.2. Carrier Transport Mechanism

Fig.3 shows J-V characteristic curves for typical (n)nc-Si:H/(p)c-Si heterojunction measured at different temperatures within the range of 230-420k. As shown in Fig.3, the measured forward current has two distinct regions. For low bias of V_F<0.8V, the current increases exponentially with applied voltage, while for higher bias of V_F>1.0V, the current increases in proportion to the power of the voltage. For reverse bias voltage, the current almost doesn't change with bias voltage, and reverse current is only on the order of nA at temperature of 230K.

In the paper, a detailed analysis of carrier transport mechanism was developed, and the main results are as follows. At forward bias voltage, the carrier transport is determined by recombination tunneling model. However, at low bias voltage (V_F<0.8V), the current is mainly determined by recombination at the (n)nc-Si:H side of the space charge region, and the slope of log J-V currents is related to temperatures; while at higher bias voltage (V_F>1.0V), the current is mainly dominated by tunneling current through sharp energy barrier ΔE_c, and the slope of log J-V currents is independent of the temperature. At reverse bias voltage, the current is determined by minority carrier tunneling in the space charge region, and doesn't change with reverse bias voltage.

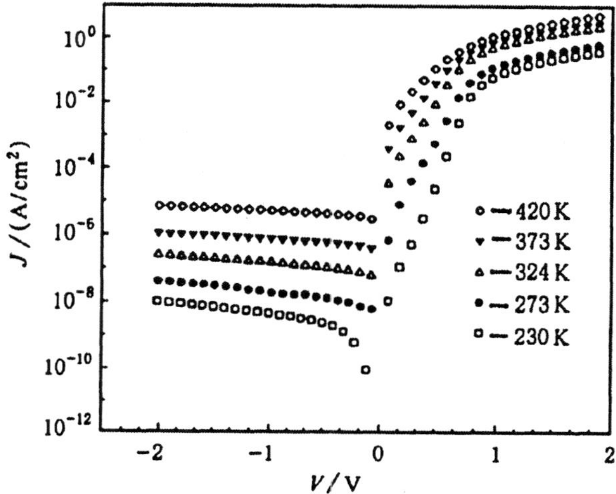

Figure 3: Current density-voltage characteristics of a (n)nc-Si:H/(p)c-Si diode measured at different temperatures

1) Low bias Voltage (V_F<0.8V)

Fig.4 shows an actual energy band diagram of (n)nc-Si:H/(p)c-Si heterojunction. It can be seen that the interface states exist in the space charge region, which can become effective recombination centers. These result from the lattice defects between (n)nc-Si:H films and (p)c-Si substrate and from the dangling bonds between nanocrystalline silicon grains and disordered interface in the (n)nc-Si:H film. Therefore, at low bias voltage (V_F<0.8V), the electrons in (n)nc-Si:H film and the holes in (p)c-Si substrate can be formed recombination current through these interface states, as shown in path B of Fig.4.

According to the usual junction rectification models, the relation between the current density and the applied voltage can be written as [10]

$$J = J_0[\exp(AV)-1] \tag{1}$$

Where J_0 is the saturation current density and A is a coefficient, which is generally dependent on the temperature. The particular expressions for A and J_0 depend on the current dominating mechanisms.

When the current is due to recombination within the depletion region, A is given by

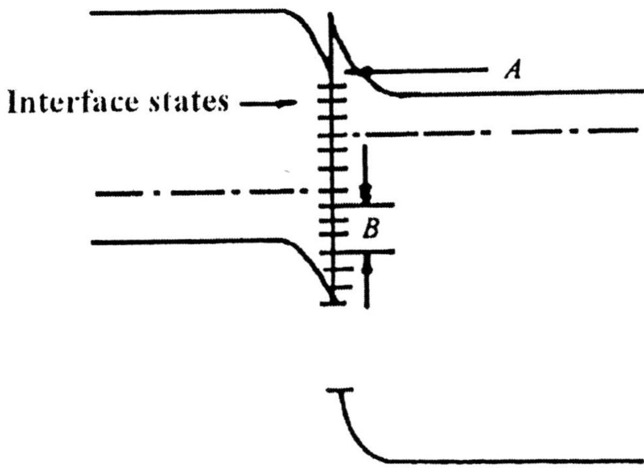

Figure 4: Energy band diagram of an actual (n)nc-Si:H/(p)c-Si heterojunction

$$A = \frac{q}{\eta KT} \quad (\eta = 1-2)$$
(2)

Where q is the electron charge; η is the ideality factor; and K is Boltzmann's constant, T is the absolute temperature. On the other hand, the temperature dependence of J_0 is

$$J_o = B \exp(^{qV_D}\!/_{KT})$$
(3)

Where B is a parameter related to characteristics of the heterojunction; qV_D is contact potential difference of the heterojunction. Finally, the relation between the recombination current density J_r through interface states and the applied voltage V is given by following expression

$$J_r = B \exp(^{-qV_D}\!/_{KT}) \left[\exp\left(\frac{q}{\eta KT}\right) - 1 \right]$$
(4)

As can be seen in the relationship (4), the recombination current is directly dependence on temperature, i.e., and is exponentially increased with increasing temperature. In fact, the dependence of recombination current density J_r on temperature mainly comes from A. As shown in Fig.5, the A values have a linear dependence on 1/KT. In addition, from the temperature dependence of A and using expression (2), the η value can be determined, which is about 1.6.

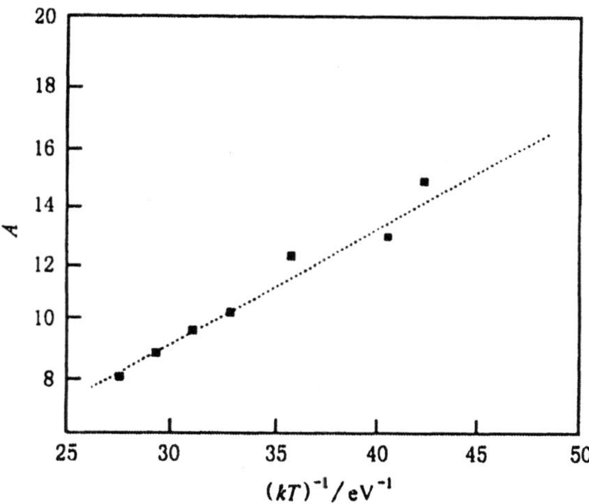

Figure 5: Temperature dependence of A

2) Higher bias voltage (V_F>1.0V)

For applied voltage of over 1.0V the carrier transport mechanism is determined by electron tunneling model. This is due to the fact that electrons in (n)nc-Si:H film have enough energy, can be through sharp energy barrier ΔE_c in the heterojunction, and reach the (p)c-Si substrate with increasing applied voltage. According to the heteroquantum dot (HQD) tunneling model, in the presence of an applied field, activated electrons have ballistic transport characteristics in nanocrystalline silicon grains and quantum tunneling occurs through the interface barrier in the (n)nc-Si:H side[11-12]. The A path in Fig.4 shows tunneling electron current through sharp energy barrier ΔE_c of the heterojunction.

The tunneling current is determined by effective electron concentration and tunneling probability. This relation can be written as [13]

$$J_\tau \approx n_e p \tag{5}$$

Where n_e is effective electron concentration; P is electron tunneling probability, and P can be given by

$$P \approx C \exp[-(\frac{16\pi}{3h})(\frac{\varepsilon_2 m_e}{N_d})^{\frac{1}{2}} \cdot (\frac{\varepsilon_1 N_a}{\varepsilon_1 N_a + \varepsilon_2 N_d})(V_D - X)] \tag{6}$$

Where C is a parameter related to characteristics of the heterojunction, m_e is the effective mass of electron ε_1 and ε_2 are dielectric constants of (p)c-Si substrate and (n)nc-Si:H film, N_a and N_d are accept and donor concentration, respectively. Thus, tunneling current is obtained by

$$J_\tau = C\, n_e \exp(\beta V)$$
(7)

Where

$$\beta = (\frac{16\pi}{3h})(\frac{\varepsilon_2 m_e}{N_d})^{1/2}(\frac{\varepsilon_1 N_a}{\varepsilon_1 N_a + \varepsilon_2 N_d})$$
(8)

As can be seen from Eq. (7) and (8), Log J-V characteristic is a typical linear relation for electron tunneling model, the slope of the curves is independent of the temperature, i.e., log J-V curves are same at different temperatures, as shown in Fig.3.

3) Reverse Bias Voltage

When the (n)nc-Si:H/(p)c-Si heterojunction was reverse biased, current density almost doesn't change with applied voltage at the temperature range of 230-420K, and reverse currents are very small, which as low as 10^{-9}-10^{-7}A. As seen in Fig.3, the log J-V curves are parallel to reverse voltage axis. It can be verified that the carrier transport mechanism is the electron tunneling model.

4   Temperature Properties of the Heterojunction

As stated above, the (n)nc-Si:H/(p)c-Si heterojunction has well electrical properties than a-Si/c-Si and ps-Si/c-Si heterojunctions. It is clearly shown in Fig.3 that the forward and reverse current densities are increased 1.5 order of magnitude when the temperatures are increased from 230K to 420K, while the forward and reverse current densities are increased an order of magnitude when the temperatures are increased from 309K to 394K. This directly verifies that our heterojunction diode has good temperature stability. Furthermore, Fig.6 shows I-V characteristics of a typical (n)nc-Si:H/(p)c-Si heterojunction diode. It is known that the heterojunction diode not only has good temperature stability, but also has higher reverse breakdown voltage, which is as high as -75V. The main reasons of the good temperature stability are as follows. First, the fabricated (n)nc-Si:H film has high conductivity ($\sigma = 10^{-1}$-$10^{1}\,\Omega^{-1}.cm^{-1}$) and small activation energy (E_{ac}=0.1-0.2eV) so that both the single electron transport in (n)nc-Si:H film and the electron tunneling through sharp energy barrier ΔE_c in the heterojunction can occur. Second, since (n)nc-Si:H film has a optical energy gap of 1.70-1.85eV, it can markedly reduce the intrinsic carrier excitations in the (n)nc-Si:H film. Third, the (n)nc-Si:H/(p)c-Si heterojunction is a typical abrupt diode structure, without the amorphous buffer layer between the (p)c-Si substrate and the thin (n)nc-Si:H film, so that it can improve interface properties of the heterojunction.

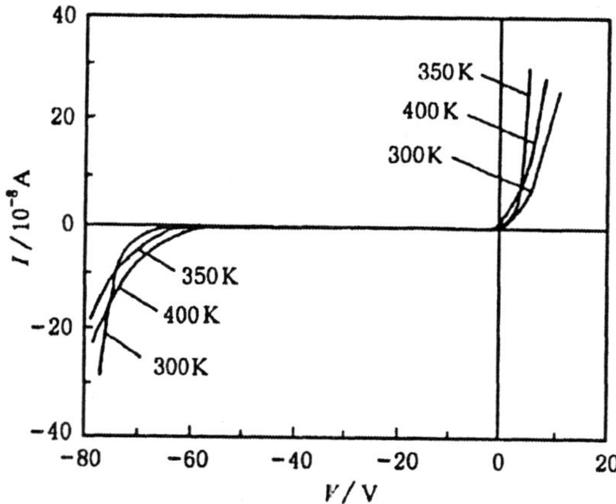

Figure 6: I-V characteristics of a typical (n)nc-Si:H/(p)c-Si diode structure.

5 Conclusions

The (n)nc-Si:h film was deposited by using conventional PECVD method, and fabricated the (n)nc-Si:H/(p)c-Si heterojunction diodes. We presented the energy band model of the heterojunction at different bias voltages. The I-V characteristics were measured at 230-420K temperatures. It is shown that the heterojunction has good temperature stability and abrupt properties. The carrier transport mechanism was qualitatively analyzed, and it indicated that the carrier transport mechanism is tunneling-recombination model at forward bias voltage. At low bias voltage ($V_F<0.8V$), the current is determined by recombination at the (n)nc-Si:H side of the space charge region, while the current becomes tunneling at higher bias voltage ($V_F>1.0V$). In addition, the heterojunction has good temperature stability and high reverse breakdown voltage (>-75V), and has a potential application in the new electronic devices.

References

[1]   Ping.Li, Q.Li, and C.Ander T.Salama, IEEE Trans. Electron Device ED-41 (1994) 932

[2]   Y.C.Peng, M.Liu, M.B.Yu, and Y.L.He, Chinese J. Semiconductors, 19(1998) 583

[3]   F.A.Rubinelli, J.K.Rath, and R.E.I.Schropp, J.Appl.Phys.,89(2001)4010

[4]   L.Pavesi, R.Guardini, and P.Bellutti, Thin Solid Films, 297(1997) 1736

[5]   L.F.Marsal, J.Pallare, X.Correig, J.Calderer, and R.Alcubilla, J.Appl.Phys., 79(1996) 8493

[6]   K.L.Narasimha, Phil.Mag., B77 (1998) 75

[7]   Y.L.He, C.Z.Yin, G.X.Cheng, L.C.Wang, and X.N.Liu, J.Appl.Phys., 75(1994) 797

[8]   Y.C.Peng, M.Liu, Y.L.He, and Y.X.Li, Chinese J. Semiconductors, 21(2000) 308

[9]   C.Delerue, G.Allan, M.Lannoo, and J.Lumin., 80(1999) 65

[10]   L.S.Yu, Semiconductor Heterojunction Physics, Chap. 4, Science Press, Beijing, 1990

[11]   Y.L.He, G.Y.Hu, M.B.Yu, M.Liu, J.L.Wang, and G.Y.Xu, Phys.Rev., B59(1999)15352

[12]   Y.L.He, M.B.Yu, G.Y.Hu, and M.Liu, Acta Physica Sinica, 46(1997) 1836

[13]   L.X.Ye, Semiconductor Physics (the First Volume), Higher Education Publishers, Beijing, 1983

Multicomponent Active Media Use in Solid-State and Dye Lasers with Passive Mode Locking

Alexander A. Sokolov

Abstract

The passive mode locking in a solid-state and dye lasers with a multicomponent active media is investigated numerically. The novel methods are proposed for the control of the envelope of a train of pulses generated in a solid-state lasers, and for the control of the shape of a pulse generated in dye lasers.

1　Introduction

Multicomponent media in the form of organic dyes are used as passive switches in solid-state lasers [1] in order to optimise the characteristics of a radiation pulse. The maximum pulse shortening requires the use of absorbers with short relaxation times, but such absorbers have high saturation intensities and correspondingly too high mode-locking thresholds. On the other hand, if combination of slow- and fast-relaxing absorbers are used in passive filter, the former (characterised by a low saturation intensity) can act effectively during the initial stage of the transient process selecting a single peak from the laser radiation distributed over cavity, whereas the latter absorbers form an ultrashort pulse from this peak.

We shall propose a method for the control of the envelope of a pulse train based on the use of a multicomponent active medium in a passively mode-locked solid state laser. If the active components have different saturation energies (which is readily achieved by focusing or defocusing of a light beam), they do not saturate simultaneously when pulses in a train cross them, i.e. the contributions of the various components to the amplification of these pulses are felt in very different parts of the train. This should obviously affect the train envelope and make it possible to control it by altering the number of components, their gains, and the saturation energies. We shall refer to the components with a relatively high saturation energy as 'hard' and whose with a relatively low saturation energy as 'soft' [1].

The control of the envelope of the pulse train is needed for one of the most important applications of the mode-locked solid state lasers, that is the synchronous pumping of dye lasers. In this application the envelope of a finite train of the pump pulses should be nearly rectangular. Various forms of negative feedback and what is known as the dynamic control of the cavity Q-factor are currently used to control the envelope of the pulse train. These methods made it possible to lengthen an output train to several thousands of pulses and give it a near-rectangular envelope [2]. The shortcomings of such methods include their technical complexity.

We shall investigate the possibilities of a multicomponent active medium use in passively mode-locked dye lasers, too. It is shown, that if the fast-relaxing (respect to the radiation pulse duration) passive filter is used, then a periodical modulation of the energy and the duration of a pulse takes place. Probably, such generation regime is usable as a method of the pulse shape control. One would notice, contemporary methods of the picosecond pulse shaping are based mostly on the inner-pulse phase modulation due to the non-linearity and dispersion of the laser elements material [3,4].

2 Model

We shall consider a model temporal representation of the radiation field in a unidirectional ring laser with passive mode locking and with simultaneous broadening of the gain and absorption lines. For both the active medium and the passive filter we shall use the approximation of the "thin" amplifying or absorbing medium, convenient for numerical calculations [5] and consisting of splitting a medium into several cells with weak amplification or absorption on each of them; described by the following system of equations:

$$E'(t) = E(t) + \alpha P(t)/2 + \eta(t), \tag{1}$$

$$dP/dt = -\gamma_2 P + ED, \tag{2}$$

$$dD/dt = \gamma_1(1-D) - \dot{E}P \tag{3}$$

Here, $E'(t)$ is the "slow" (varying slowly compared with the period of the optical oscillations) amplitude of the radiation field at the exit from the cell; $E(t)$ is the field amplitude at the entry of the cell; $P(t)$ is the "slow" amplitude of the polarisation induced by the input pulse; α is the gain (a positive value) or absorption (a negative value) in such a "thin" cell (in our case, $|\alpha| \leq 0.1$); $\eta(t)$ is a random function (with a correlation time which is reciprocal of the width of the gain line), modelling spontaneous emission; D is the normalised (to unity) difference between the populations of active levels of the medium; γ_2 is the rate of relaxation of the polarisation; γ_1 is the rate of recovery of the difference between the populations of the active levels of the nonlinear medium during continuous pumping. Expression (1) is obtained from the equation

$$dE/dz + (1/c)(dE/dt') = \alpha P/2, \tag{4}$$

if in this equation we ignore the spatial derivative and adopt a system of coordinates associated specifically with a moving pulse ($t=t'-z/c$) and add the term $\eta(t)$. Eqns (2) and (3) follow directly from the equations for the density matrix of a two-level quantum system with homogeneous broadening and relaxation under conditions of zero detuning of the central transition frequency from the eigenfrequency of the cavity [6].

This model will be used for the numerical investigation of both solid-state and dye lasers with the multicomponent active media.

3 Solid-State Laser. Control of the Envelope of a Train of Pulses

We shall assume that the relaxation constants of all the components of the active medium are $\gamma_2=10^{11}$Hz and $\gamma_1=10^4$Hz, which are approximately equal to the corresponding parameters of an Nd:YAG crystal. A change in the saturation energies of the active-medium components by focusing or defocusing of the light beam in each of them is taken into account by multiplying the amplitude E and $\rho_j=(Q_j s/Q s_j)^{1/2}$, where Q_j and s_j are, respectively, the cross section of a stimulated transition in a particular component and the cross-sectional area of the light beam in the component; Q and s are the corresponding values of the parameters of the component used in the normalisation procedure [7]. It is assumed that $1\leq\rho_j\leq15$ for the active medium components and that the dimensionless term is $\eta(t)\sim10^{-9}$.

Transformation of the radiation field in a fast-response passive switch is described by similar equations with $\gamma_2=10^{13}$Hz and $\gamma_1=10^{12}$Hz, and with the field amplitude multiplied by the factor $\rho_f=250\div450$ (similar to ρ_j) for the active-medium components).This corresponds to the characteristics of fast-relaxing pyrylocyanide dyes [8]. The linear losses in the cavity are assumed to be lumped in one of the mirrors with the amplitude reflection coefficient R=0.9. The round-trip time of the cavity T_r is assumed to be 5 ns. The gain A_j acquired in the jth component of the active medium and the absorption coefficient of the passive switch A_f (a negative value) used later do not refer to a "thin" layer of a specific nonlinear medium but to the medium as a whole. The differences between the population of their active levels are denoted by D_j and D_f, respectively.

The following numerical results are obtained. If the active medium contains one component (with the unsaturated gain A_1 and the population difference D_1), the input radiation is a periodic sequence of finite pulse trains (with one pulse per round trip through the cavity). The train follow with a period of several hundreds of T_r, they consist of several tens of pulses, and have a dome-shaped envelope with the leading and trailing edges of the same duration (see Fig.1a).

The main repetition period is governed by the rate of recovery of the $D_1=D_{10}$ inversion in the active medium, sufficient for bleaching the passive switch. In other words, formation of a stimulated radiation pulse from the noise background requires that the gain A_1D_{10} in the active medium should exceed the total losses in the fast-relaxing passive switch ($-A_f$)and in the cavity mirrors $G=1-R^2$ (in the case under discussion, we have $G\approx0.2$):

$$A_1D_{10} \geq -A_f + G. \tag{5}$$

It should be noted that an increase in ρ_f, i.e. an increase in the absorber "softness" relative to the active element, reduces the number of pulses N in a train, whereas the maximum dimensionless energy ε_m of a pulse in a train and the integral dimensionless energy W of the train both increase. A quantity similar result is obtained by increasing the passive switch density A_f. For example, if $A_1=0.9$, $\rho_1=1$, $\eta(t)\approx10^{-9}$, $T_r=5$ns, $R=0.9$, it is found that $N=77$, $W=17.6$, $\varepsilon_m=1.1$ for $A_f=-0.24$, $\rho_f=250$ (the train envelope is shown in Fig.1a); we correspondingly obtain $N=35$, $W=24.4$, $\varepsilon_m=2.6$ for $A_f=-0.35$, $\rho_f=250$ and $N=46$, $W=29.8$, $\varepsilon_m=2.4$ for $A_f=-0.24$, $\rho_f=450$ (the train envelope is shown in Fig.1b). We can see that the train is shorter and "steeper" than that in Fig.1a. The number of pulses N in the train is found on the assumption that the minimum energy of a pulse in a train is not more than 300 times less than the maximum energy ε_m of a pulse in the train.

It should be also noted that the energy of a pulse in a train increases, i.e. the leading edge of the train is formed when the gain A_1D_1 of a gradually saturating active element still exceeds the total radiation losses in the mirrors and in the passive switch. At the moment when a pulse in a train reaches its maximum energy ε_m, the gain A_1D_1 in the active element falls to the gain A_1D_{11}, approximately equal to the mirror losses (the losses in the passive switch are ignored because the switch is strongly bleached):

$$A_1D_{11} \approx G. \tag{6}$$

Further reduction in the inversion in the active element begins to reduce the energy of a pulse in the train, i.e. the trailing edge of the train envelope is being shaped. It is clear from expression (6) that when the losses in the mirrors are sufficiently low ($G\approx0.2$), the gain A_1D_1 in the active element up to the moment of attainment of the maximum energy ε_m of a pulse is also fairly low, i.e. the active elementy can be regarded approximately as fully saturated.

Let us assume that a second amplifying component is now added to the active medium, that this new component is "harder" (parameter $\rho_2=0.1$) than the first, and that it is characterised by a low (compared with the first component) unsaturated gain which is approximately equal to the mirror losses:

$$A_2 \approx G < A_1. \tag{7}$$

Let us assume, for the sece of simplicity, that the new component has the some relaxation constants as before: $\gamma_2=10^{11}$Hz, $\gamma_1=2\times10^4$Hz. For a system of this kind, in view of the smallness of the gain in the second component, the condition for the onset of shaping of the next train retains-in the first approximation-its previous form [see expression (5)]:

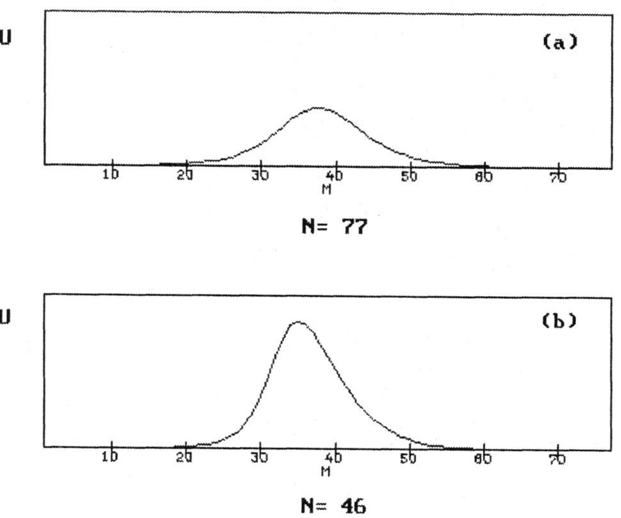

Figure 1: Dependences of the energy U of pulses in a train on the serial number of a pulse M calculated for $A_1=0.9$, $\rho_1=1$, $A_f=-0.24$, $\rho_f=250$, $\eta(t)\approx10^{-9}$, $T_r=5$ ns, $R=0.9$, when the number of pulses in a train is $N=77$ (a) and 46 (b).

$$(A_1+A_2)D_{10} \approx A_1D_{10} = -A_f + G. \tag{8}$$

Next, when the energy of a pulse in a train approaches ε_m (which is the same as in the case of a single component active medium discussed above) and the "soft"component of the active medium can be regarded as saturated, the "hard"component of the medium is not yet saturated (specifically because of its "hardness"). Then, if the gain acquired in the second component (when the inversion in this component is $D_{20}=D_{10}$) compensates the energy losses experienced by the pulse in the mirrors and in the passive switch,

$$A_2D_{20} = G, \tag{9}$$

the energy of a pulse in the train envelope becomes flattened. In the roughest approximation, if the second component has a saturation energy k times greater, the upper part of the train envelope "stretches" by the factor k. A train with this envelope is shown in Fig.2. Its flat part, where the energy of a pulse does not vary by more than 10%, contains $N_p=37$ pulses.

Since pumping of the active medium components in the course of passage of a pulse train though them can be ignored, the difference between the saturation energies of the components is governed only by the ratio of their factors ρ_j. Naturally, the higher this ratio, the slower is the saturation of the second component and the flatter the envelope of the train. Our numerical results show that, if $\rho_1/\rho_2 \approx 10 \div 15$, it is possible to generate trains in which several tens of pulses have approximately the same (differing by no more than 10%) energy. In order to "prolong"the action of the "hard"component, we shall assume that a third (absorbing) component is introduced into the active medium and that its "hardness"is intermediate between the two amplifying components and that it satisfies the condition

$$A_2 D_{20} + A_3 D_{30} = G, \tag{10}$$

$$A_2 D_{20}\rho_2{}^2 = - A_3 D_{30}\rho_3{}^2. \tag{11}$$

Here, A_3 is unsaturated gain of the third component; D_{30} is the difference between the populations of its active levels up to the moment of attainment of the maximum energy ε_m of a pulse in the train. Relationship (10) means that, after saturation of the "soft"amplifying component, the total gain of the remaining two components is equal to the losses of the energy of a pulse in a single trip through the cavity, whereas relationship (11) means that further saturation of the "hard"amplifying component is compensated- in the first approximation- by bleaching of the absorbing component, i.e. the combined effect of the two remains unchanged. Our numerical results indicate that such an operation makes it possible to lenghten by about 25% the part of the train with a constant pulse energy.

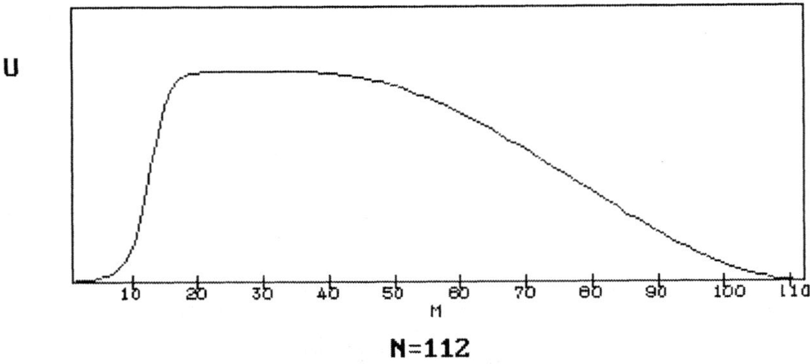

Figure 2: Dependences of the energy U of pulses in a train on the serial number of a pulse M calculated for $A_1=0.84$, $\rho_1=1.0$, $A_2=0.0325$, $\rho_2=0.07$, $A_f=-0.5$, $\rho_f=250$, $\eta(t)\approx10^{-9}$, $T_r=5$ ns, $R=0.9$, $N=112$.

In the case of a solid-state lasers generating finite pulse trains it is fundamental that, because of the probabilistic mechanism of the formation of a pulse from fluctuations of the spontaneous radiation and because of variation in the pump parameters, the train characteristics are reproduced only very approximately. An estimate of the influence of these effects in the equations adopted in our model were obtained by altering γ_1 and $\eta(t)$. It was found that, for example, when the parameters correspond to whose in Fig.2, a reduction or an increase in $\eta(t)$ by an order of magnitude does not affect the characteristic flat top of the train, but it alerts the total energy W of the train within 10%, changes the length of the flat part within 16%, and leaves practically unchanged the maximum energy ε_m of a pulse in the train. However, the train envelope is much more sensetive to a change in the pumping rate. When other parameters are unaffected, a train retains its flat top only in the range $\gamma_1=(1.95\div3)\times10^4$Hz and in this range its total energy W changes approximately twofold, the number of pulses N in the train changes by 50% , whereas the parameters N_p and ε_m are only slightly affected.

The effectiveness of the proposed method is undoubtely less than the method of feedback and control of the cavity Q-factor, which makes it possible to generate sequences of several thousands of low-energy pulses under quasy-cw conditions. Our results is the generation of trains of several tens of approximately the same pulses when the pumping during the growth of the train can be ignored and it resembles very closely what is known as the finite train regime in solid-state lasers with negative feedback and controlled cavity Q-factors [9]. However, in our case this result is achieved much more simply in the technical sence without recoure to negative feedback devices.

This simplicity may possibly make it desirable to use milticomponent active media in lasers with negative feedback and a controlled cavity Q-factor as an auxiliary means for the control of the train profile. A multicomponent active medium may be used as an independent means of control in synchronous dye laser systems with hard excitation (for example, if an ultrathin cavity is imployed [10]) , when a pump train of length amounting to several tens of pulses is quite sufficient for quasy-cw lasing.

4   Dye Laser. Modulation of the Energy and Duration and Control of the Shape of a Pulse

We shall assume that the active medium consists of two separated components with the relaxation constants of both them $\gamma_2=10^{13}$Hz and $\gamma_1=10^9$Hz and the parameters of "softness" $1\le \rho_j\le5$, j=1,2. It is assumed that $T_r=5$ ns, R=0.9, $\eta(t)\sim10^{-9}$.

According to our numerical results, a multicomponent active medium use involves the periodical modulation of the energy and duration of a radiation pulse (with the period of modulation equal to some tens cavity round trip periods T_r; see Fig.3), whereas for a single-element active medium it takes place a "stationary" mode locking regime with the constant pulse characteristics. The more difference between the saturation energies of the active medium components (i.e. the difference between their parameters ρ_j), the deeper modulation is (see Fig.4, 5).Principal conditions of the modulation regime are the small (respect to the radiation pulse duration) relaxation time of the passive switch, not more than 1-3 ps, and its sufficient absorption A_f. When increasing $|A_f|$ gradually, first is achieved a threshold of the stationary mode locking regime (one pulse with the constant characteristics per the cavity round trip); only then it is achieved a threshold of the modulation, above which the amplitudes of the pulse energy and duration increase about linearly versus $|A_f|$ (see Fig.6).The energy and the duration of a radiation pulse oscillate with about 180° phase shift one respect the other; the maximum (minimum) of the pulse duration is achieved some cavity round trips before the minimum (maximum) of the pulse energy. The shape of a pulse periodically changes, too (shown in Fig.7).

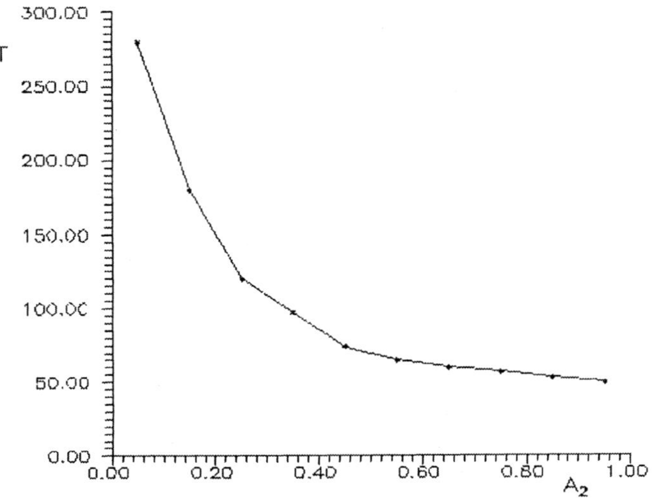

Figure 3: Dependences of the modulation period T (in the number of the cavity round-trip times T_r) on the gain A_2 of a "soft" amplifying component. $A_1=0.55$, $\rho_1=1.0$, $\rho_2=5.0$, $A_f=-0.25$, $\rho_f=5.0$.

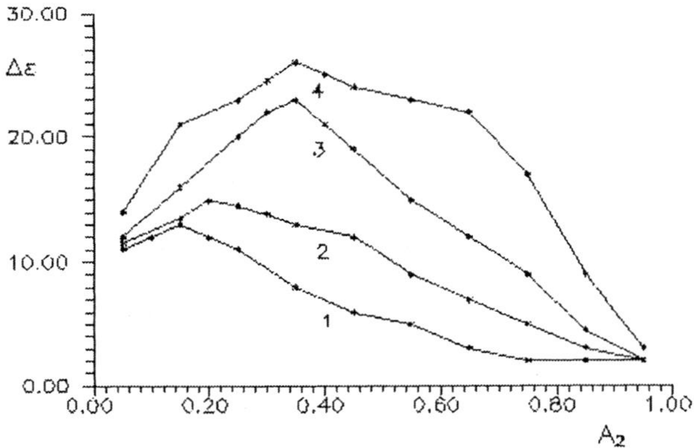

Figure 4: Dependences of the amplitude $\Delta\varepsilon$ of the pulse energy oscillations (in percents of an average energy value) on the gain A_2 of a "soft" amplifying component. $A_1=0.55$, $\rho_1=1.0$, $\rho_2=1.4$ (curve 1), $\rho_2=2.0$ (curve 2), $\rho_2=3.0$ (curve 3), $\rho_2=5.0$ (curve 4), $A_f=-0.25$, $\rho_f=5.0$.

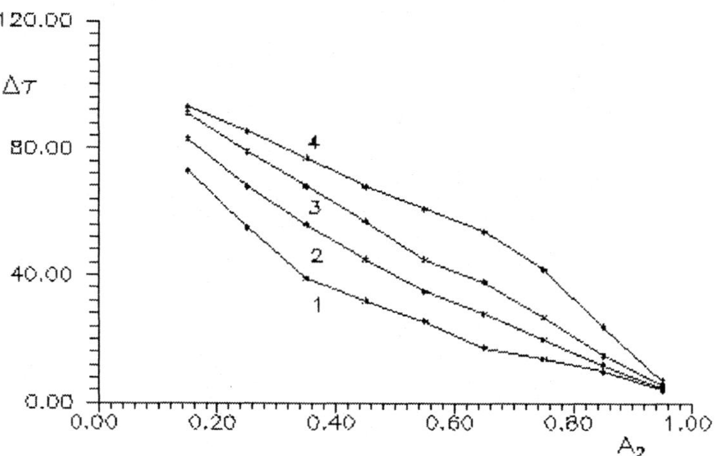

Figure 5: Dependences of the amplitude $\Delta\tau$ of the pulse duration oscillations (in percents of an average duration value) on the gain A_2 of a "soft" amplifying component. Volumes of parameters and curve marks, such as on the figure 4.

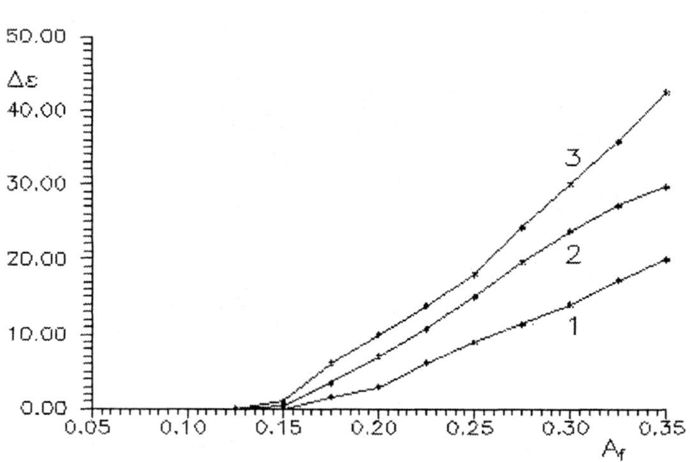

Figure 6: Dependences of the amplitude $\Delta\varepsilon$ of the pulse energy oscillations (in percents of an average energy value) on the absolute value of the absorption $|A_f|$ of a passive switch. Volumes of parameters and curve marks, such as on the figure 4.

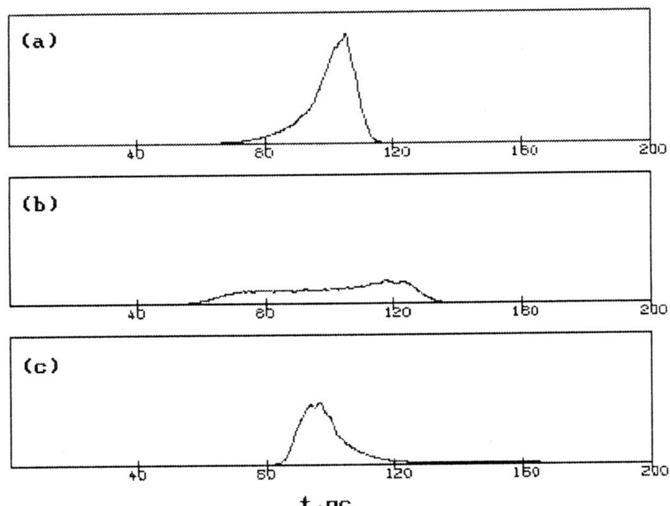

Figure 7: The consecutive deformation of a pulse intensity profile (the sequence of the profiles (a), (b), (c) is taken with a "step" of the pulse 20 cavity round trips). A_1=0.55, ρ_1=1.0, A_2=0.4, ρ_2=5.0, A_f=-0.25, ρ_f=5.0.
The X-axis is time (in picoseconds).

We propose an easy explanation for these results. If an active medium consists of some components, the "soft" one amplifyes the leading edge of the pulse mostly, whereas a fast- relaxing passive filter absorbs the energy on the trailing edge of the pulse. These deforms the shape of the pulse and makes its duration change (see Fig.7). Since the action of a passive switch depends on the pulse intensity, when having different shapes and duration, the pulse has different losses and therefore attains different energies. Naturally, the more difference between the saturation energies of the active elements is (i.e. difference between ρ_1 and ρ_2), the stronger modulation is.

Looking at Fig.3-5, one can see that the more "softness" ρ_2 and the less gain A_2 of the "soft" component of an active medium are, the more period and amplitudes of the modulation of a pulse energy and duration are. It is easily explainable, too. Increasing ρ_2 and (or) decreasing A_2 makes the action of the "soft" component decrease and "shift" to the leading edge of a pulse. Therefore the deformation of a pulse slows down and changes its view. For example, for ρ_2=5.0 and A_2 0.8 the profile of a pulse is now wide and low, now narrow and high (preserving its dome-like shape); for 0.3 A_2 0.8 a pulse is now dome-shaped, now about rectangular-shaped (see Fig.7); for 0.1 A_2 0.3 a pulse becomes two-domed, and its deformation means the growth of the leading dome while the trailing dome decays; at last, for A_2 0.1 a pulse devides into two separated pulses, the leading one grows and the trailing one decays gradually.

We propose an analitical interpretation of these numerical results, based on the following balance equations:

$$\Delta W_j = \int_0^W a_j(w)\, dw; \tag{12}$$

$$a_j(w) = A_j \exp(-\sigma_j w). \tag{13}$$

Here, ΔW_j an increase of the energy of a pulse having the initial energy W after passing the jth active component having the initial gain A_j ; $a_j(w)$ is the gain of the jth component as a function of the energy w of radiation passed through it (the relaxation process is ignored), $\sigma_j = \rho_j^2$. We shall consider a case of the sufficiently small A_2 and sufficiently high ρ_2, when the radiation pulse devides into two separated pulses during the modulation process.

We determine the energy W of a pulse (not considering its modulation) from that its increase ΔW per a cavity round trip is equal to zero:

$$\Delta W = \int_0^W (A_1 \exp(-\sigma_1 w) + A_2 \exp(-\sigma_2 w) + a_f(W,\tau) - G)\, dw = 0. \tag{14}$$

Here, $G = 1 - R^2$ is the linear losses factor, $a_f(W,\tau)$ is some "effecting" saturated absorption of the passive filter for a pulse with the energy W and some duration τ. In the first approximation [proposing G small and ignoring $a_f(W,\tau)$ in (14)] the energy W of a pulse is

$$W \cong (A_1/\sigma_1 + A_2/\sigma_2) / G. \tag{15}$$

To calculate a period of modulation we shall consider a case of the small A_2, when a radiation pulse is devided into two separated pulses. It is assumed the following scheme of the modulation process. Due to the amplification of the leading edge of a pulse by a "soft" active element, a new pulse arises there. The energy growth ΔW_2 of the new pulse per a cavity round trip (proposing the full saturation of the "soft" amplifying component and ignoring the action of the "hard" component) is

$$\Delta W_2 \cong \int_0^{+\infty} A_2 \exp(-\sigma_2 w)\, dw = A_2/\sigma_2. \tag{16}$$

Then the energy $W_2(n)$ of the new pulse on the nth cavity round trip (ignoring both linear G and nonlinear a_f losses) is:

$$W_2(n) = n\, \Delta W_2 = n\, A_2 / \sigma_2. \qquad (17)$$

We shall estimate a period of the modulation as the time from the new front pulse arise till it attain the maximum energy W (15), when the "old" rear pulse has decayed. This period of modulation T (in a number of the round trip times T_r) is

$$T = W/\,\Delta W_2 = (A_1/\sigma_1 + A_2/\sigma_2)\,/\,(G\,A_2/\sigma_2). \qquad (18)$$

After many approximations used, the results of calculations on expressions (15), (18) seem to correspond only qualitatively to the numerical results. So, for example, for $A_1=A_2=0.55$, $\rho_1=1.0$, $\rho_2=5.0$,

$A_f=-0.25$, $\rho_f=5.0$, $G=0.19$ the energy W of a pulse is 3.0, when calculated on expression (15); for the same parameters, the numerical result is W=1.2.The period of modulation, when calculated for the different values of parameters on (18), differs also by 200-300% from a corresponding numerical result. However, we propose eqns (15), (18) being usable for quantitative estimates.

5 Conclusions

The numerical results obtained show, that a multicomponent active medium use in a passively mode-locked solid-state or dye laser changes essentially the view of the mode locking and makes it possible to control its features, i.e. the shape of a pulse (in the dye lasers) and the envelope of a pulse train (in the solid-state lasers).

One would take into account some analogy between a pulse train (as a whole) in a passively mode-locked solid-state laser and a single radiation pulse in a passively mode-locked dye laser, whose characteristics are changed and controlled. In both cases, the leading edge of a pulse (train of pulses) undergoes the action of a "soft" amplifying component, and therefore changes its form. This analogy is limited by a feedback presence in the case of a solid-state laser (any pulse is a previous one after the round trip through the cavity), which makes the envelope of a pulse train stable, whereas in the case of a multicomponent dye laser the shape of a pulse is unstable and changes periodically.

References

[1] Demchuk M I, Kolchanov I G, Manichev I A, et al. Kvantovaya Electron. (Moscow) 15, 2427 (1988) [Sov. J. Quantum Electron. 18, 1523 (1988)].

[2] Zaporozhchenko V A Kvantovaya Electron. (Moscow) 21, 843 (1994) [Quantum Electron.24, 785 (1994)].

[3] Ahmanov S A, Vislouh V A, Chirkin A S Uspehy Fizicheskih Nauk, 149, 499 (1986).

[4]   Wiener A M, Heritage J P, Thurston R N Opt Lett, 13, 153 (1986).

[5]   Demchuk M I, Manichev I A, Mikhailov V P Kvantovaya Electron. (Moscow) 18,61 (1991) [Sov. J. Quantum Electron. 21, 54 (1991)].

[6]   Sargent M III, Scully M O, Lamb W E Jr Laser Physics (Addison-Wesley, Reading, MA, 1974).

[7]   New G H C Proc. IEEE 67,380 (1979).

[8]   Babenco V A, Dyadyusha G G ,Kudinova M A, et al. Kvantovaya Electron. (Moscow) 7, 1796 (1980) [Sov. J. Quantum Electron. 10, 1035 (1980)].

[9]   Bayanov I M, Biglov Z A, Gordienco V M, et al. Izv. Akad. Nauk SSSR Ser. Fiz.54, 2464 (1990).

[10] Egorov K D, Nekhaenco V A, Pershin S M, et al. Kvantovaya Electron. (Moscow) 13, 1169 (1986) [Sov. J. Quantum Elctron. 16, 768 (1986)].

Key Physical Problems in Laser Separation of Weighable Amounts of Ytterbium-168[1]

S. I. Yakovlenko

General Physics Institute of Russian Academy of Science

Vavilov street 38, V-333 QSP-1 Moscow, Russia 117942

Phone: (095)1328280   Fax: (095)1357922   E-mail: syakov@kapella.gpi.ru

Abstract

A review of the AVLIS (atomic vapor laser isotope separation) development performed by General Physics Institute (Russian Academy of Science) and LAD Ltd (Moscow, Russia) during 1994-1998 years is presented. The processes of laser isotope separation of Yb have been studied analytically, with computer simulations and experimentally. The basic attention is given to following topics: the selectivity of ionization; the lasers beams arrangement in a cavity; the forming of a vapor flux; the extraction of ions from plasma. The key problems of laser separation of weighable amounts of rare isotope were solved. Installations to produce highly enriched ^{168}Yb in industrial scales were created. The contents of ^{168}Yb in the laser produced plasma was reached 90 - 95 %, in the material deposited on the ions collector - up to 62 % and in the washing liquid - up to 45 %. The rate of the enriched ytterbium production was up to 5—10 mg/hour (over 1 g per month). For the first time a profitable commodity was produced by the AVLIS method.

Keywords: AVLIS installation; highly enriched Yb168; selective photoionisation; extraction of ions; cavity

---

[1] This paper is an updated material of the reviews: S.I. Yakovlenko, 'Principal physical problems in laser separation of weighable amounts of a rare ytterbium isotope'. *Kvantovaya Elektron. (Moscow)* **28** (11) 971-987 (1998) [*Quantum Electron.* **28** (11) 945-961 (1998)]; S.I. Yakovlenko, 'Main physical problems of AVLIS technology in producing ytterbium-168 in weighable amounts'. *Laser and Particle Beams* **16** (4) 541-568 (1998)

1. Basic problems

This review describes the current status of theoretical and experimental investigations, carried out during 1994-1998 years in Russia (at the Institute of General Physics, Russian Academy of Sciences and at the 'Lad' Scientific-Production Enterprise, Moscow). These investigations resulted in the construction of the first ever facilities [1 - 5] capable of commercially viable production of the enriched ^{168}Yb isotope by the AVLIS (atomic vapour laser isotope separation) method.

The AVLIS method, proposed by V S Letokhov [6], consists of selective multistage photoionisation of metal vapours followed by extraction of an isotopically enriched ion component from plasma.

Two main practical applications of this method were in operation:

- production of relatively large quantities (of the order of a tonne per year) of a weakly enriched product (^{235}U) for nuclear power requirements;

- production of highly enriched (with the enrichment coefficient in excess of 20%) isotopes in much smaller volumes for use in medicine.

In spite of the fact that the first of these applications has been investigated for over 25 years and that in the USA alone this investigation cost about two billion dollars, a laser technology capable of competing economically with the traditional methods is not yet available. These investigations are stopped now. However, studies of the second application have recently met with considerable success. One of the striking achievements is the development of a method for producing a highly enriched ^{168}Yb isotope in amounts quite sufficient for medical applications. The abundance of this isotope in natural mixtures is only 0.14% and practical use is made of a mixture enriched to over 20%-25%, the price of which (according to the Oak Ridge catalogue) exceeds 300 000 dollars per gramme.

The principal scheme for three-stage selective ionisation of ytterbium (Fig. 1) was proposed in 1991 [7]. A scheme for implementation of the AVLIS method, designed for efficient production of considerable amounts of the isotope, is shown in Fig. 2. In selective ionisation one can employ, as usual, the radiation from tuneable dye lasers pumped by copper vapour lasers.

However, implementation of this scheme required a solution of a number of important scientific and engineering problems, which are not encountered in laboratory demonstrations of the effect or in research. The main directions of work done on the industrial use of the AVLIS method are:

—formation of spectral and spatial laser radiation structures to ensure selective photoionisation in a relatively large volume;
—construction of a source ensuring vapour flow with the necessary properties;
—construction of a system for ion extraction from a plasma.

The decision to aim for commercially viable production of weighable amounts of ^{168}Yb was the reason for major departures from the previous research investigations and from the work on production of nuclear fuel for power reactors, as well as from several other laboratory studies. It proved possible to employ fairly small facilities with low power demands and to avoid complex problems associated with electron-beam evaporation of the material.

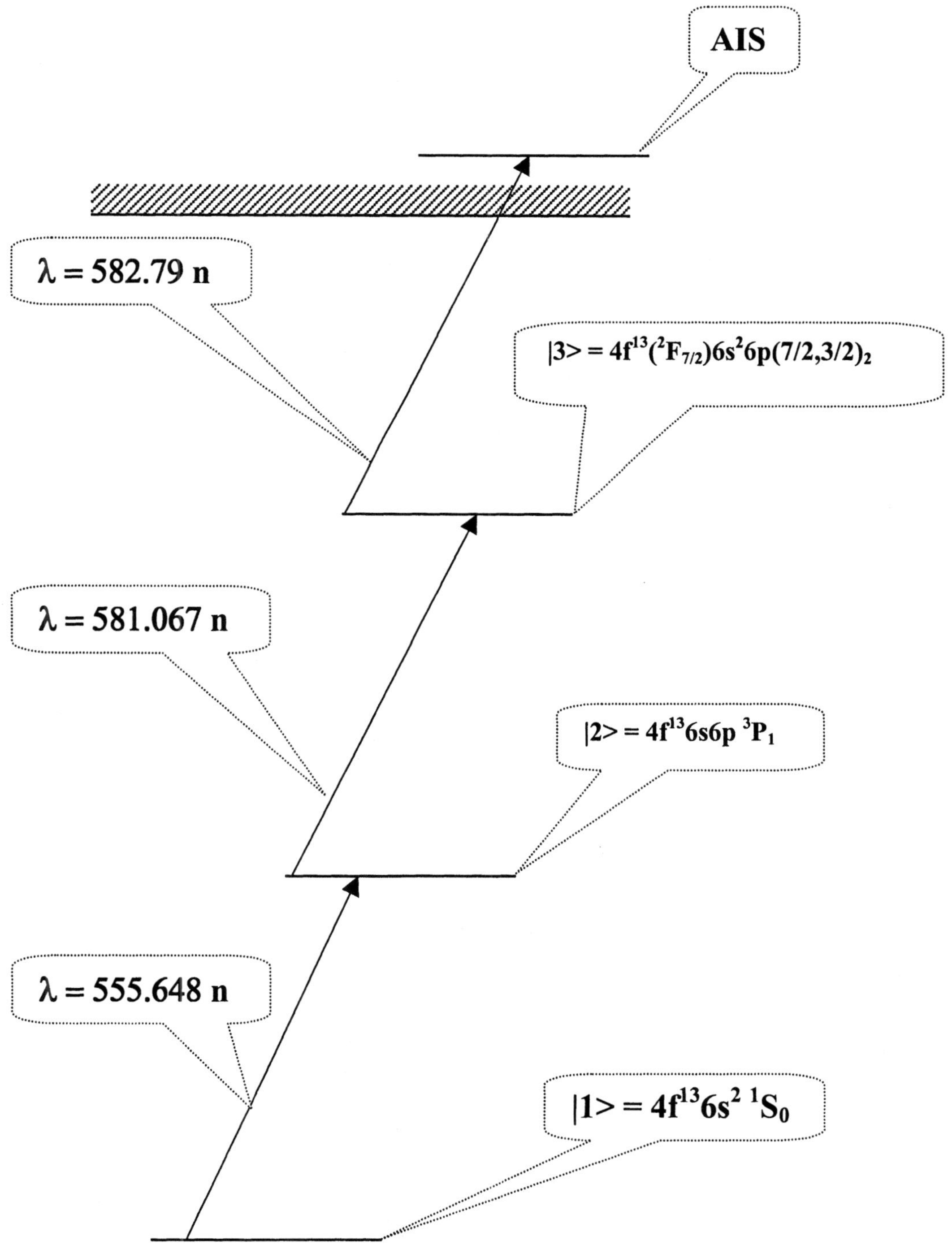

Figure 1: The scheme of 3-stage ionisation of ytterbium

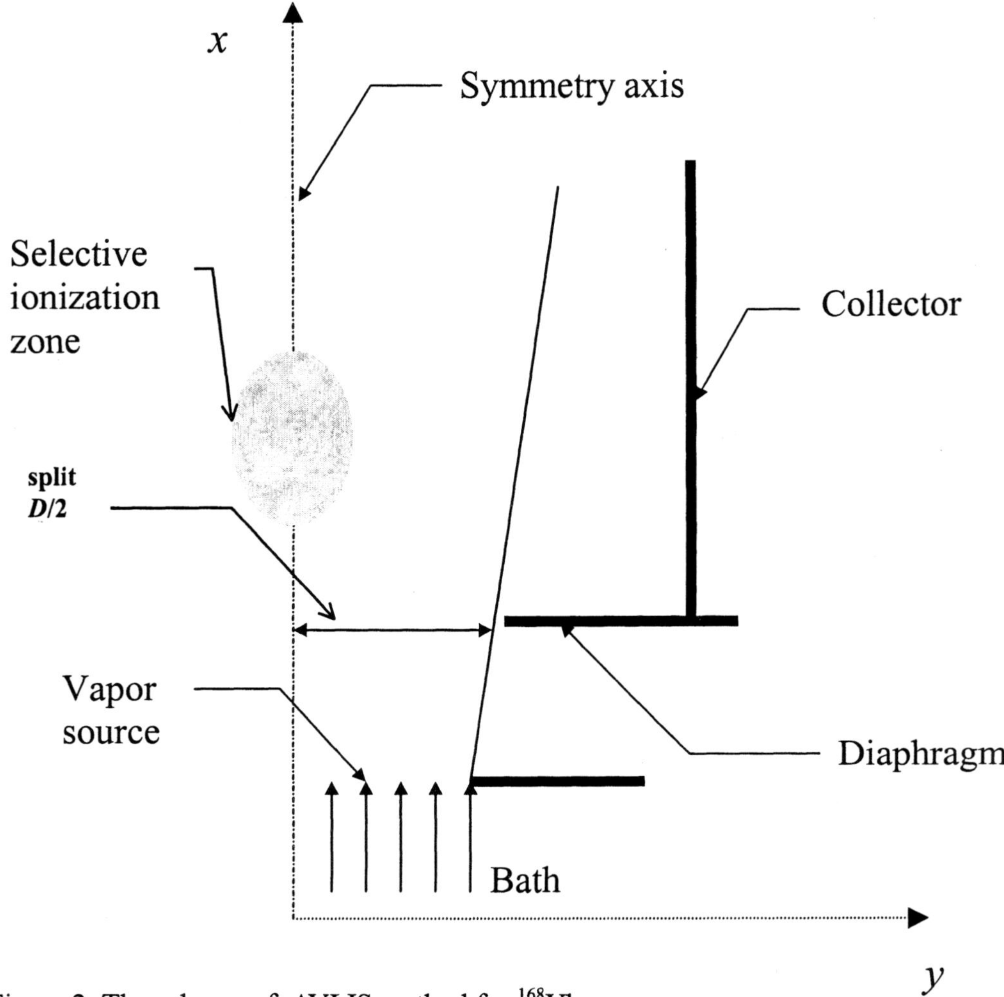

Figure 2: The scheme of AVLIS method for ^{168}Yb

However, in view of the low abundance of the target isotope in the initial mixture, much greater difficulties were encountered. These difficulties are in ensuring the selectivity and efficiency of photoionisation in a relatively large volume [5, 8-10] and also in dealing with the problem of contamination of the collector in the course of formation of vapour flow [11 - 17]. The need to produce weighable amounts of the rare isotope magnified the difficulties because of the considerable space charge that appears in the course of ion extraction [18-23]. It must be stressed particularly that these problems had to be solved jointly. For example, in constructing a system for ion extraction we could not forget the collector contamination hazard and the width of a laser radiation line had to match the Doppler width of atomic lines of a given vapour source.

We tackled these problems, relying extensively on the results of numerical simulations. Numerical simulations accompanied the manufacture of an each important element of the installations on an accessible level. Computations carried out at the Institute of General Physics of the Russian Academy of Sciences made it possible to reduce considerably the volume of preliminary experiments and the amount of work done on optimisation of the facility parameters.

Preliminary experiments were carried out on the first facility based at Troitsk (Moscow district), production of sub-gramme amounts of ^{168}Yb was achieved in the second facility [1,2] located in the 'Lad' firm in the city of Khimki (Moscow). Recently, a third facility, with a longer vacuum chamber, was put into service at Khimki [3, 4].

Presented review will deal with some scientific aspects of this series of investigations. General information on other research in the field of laser isotope separation can be obtained from reviews and summarising communications [24-39]. Characteristics of the dye lasers used in the AVLIS method are given in Refs. [40-58].

2 Photoionisation selectivity

2.1 Selective ionisation scheme

We shall consider the separation of ^{168}Yb from a natural mixture, involving the three-stage-ionisation via the autoionising state (AIS):

$$\left|4f^{14}6s^2\,{}^1S_0\right\rangle \xrightarrow{\lambda=555.648\ \text{nm}} \left|4f^{14}6s6p\,{}^3P_1\right\rangle \xrightarrow{581.067} \left|4f^{13}({}^2F_{7/2})6s^26p(7/2,3/2)_2\right\rangle \xrightarrow{582.79} \text{AIS}$$

(Fig.1). We shall further consider a linearly polarised light.

The experimental probabilities of the $1\to2$ and $2\to3$ transitions are $A_{12}=3.6\cdot10^6\cdot Q(0IFM,1IF'M',\ \mu)$; $A_{23}=1.9\cdot10^5\cdot Q(1IFM,2IF'M',\ \mu)$; Q is the angular factor for the probability of E1 transition between components of a hyperfine structure:

$$Q(JIFM, J'IF'M',\mu) = (2F+1)(2F'+1)\begin{pmatrix} F & 1 & F' \\ -M & \mu & M' \end{pmatrix}^2 \begin{Bmatrix} J & F & I \\ F & J' & 1 \end{Bmatrix}^2$$

Here (…) - is the $3jm$ Wigner symbol, {…} - the $6j$ symbol; F is the total momentum of an atom, J and I are the moments of the level and the nucleus, respectively, M is the projection of the total momentum, μ is the radiation polarisation index ($\mu=0$, for linear polarisation and $\mu=\pm1$, for circular polarisation).

The photoionisation cross section is proportional to the same angular factor and it can be converted similarly. It is necessary yet to take into account, that the photoionisation can occur via hyperfine states of AIS with different values of F:

$$\sigma = 7.8\cdot10^{-15}\sum_{F} Q(2IFM,3IF'M,\mu).$$

The line broadening usually can by regarded as the Doppler type with the spectral function

$$S_{\mathrm{D}}(\Delta\omega) = \frac{1}{\sqrt{\pi}\Delta\omega_{\mathrm{D}}}\exp[-(\Delta\omega/\Delta\omega_{\mathrm{D}})^2], \quad \Delta\omega_{\mathrm{D}} = \omega_0 v_T/c, \quad \Delta\omega = \omega - \omega_0, \qquad (1)$$

where ω_0 is the unshifted transition frequency; $v_T = \sqrt{2T/M_{\mathrm{Yb}}} \approx 1{,}1\cdot10^5\sqrt{T}$ is the thermal velocity of vapor atoms ($M_{\mathrm{Yb}} \approx 173\times1.67\ 10^{-24}$ gm, T is a gas temperature). In estimates we can assume that $T \approx 0.1$ eV and $\lambda \approx 580$ nm, we than have $v_T \approx 3.4\cdot10^4$ cm/s, and consequently $\Delta\omega_{\mathrm{D}} \approx 3.6\cdot10^9$ s^{-1}, $\Delta\nu_{\mathrm{D}} \approx 570$ MHz.

2.2 Kinetic models

Preliminary comments.

In general, a full analysis of selective photoionisation should be based on the Maxwell equations for an electromagnetic field and on the material equations for a medium, which are in turn based on kinetic equations for the density matrix. One should take into account various fine effects, including focusing and defocusing of laser beams, formation and decay of solitons, etc. Such an analysis can be found in, for example, Refs. [71 -75]. However, if use is made of multimode lasers and the fine coherent effects are ignored, it is possible to use the equations for radiative transfer in combination with the equations for the balance of populations of the atomic levels. An approach based on the radiative transfer equations has made it possible to reveal fairly fine effects [105, 106] and even to develop a non-linear theory of spectral line broadening.

Moreover, the simple approach utilising the radiation intensity concept is often more reliable than an overcomplicated analysis which takes account of the coherent effects. In the letter case it is necessary to make other simplifying assumptions about the property of the radiation and of the medium, which are quite difficult to satisfy under real conditions. For example, if the density matrix formalism is used, it is difficult to analyse real experimental conditions under which radiation from three multimode laser beams interacts in a three-dimensional cavity with multilevel absorbing atoms of many isotopes (see Section 3). The use of the radiative transfer equations is justified by the good agreement between the theoretical predictions and the experimental results.

In the present section we shall consider the kinetics of selective photoionisation ignoring propagation of a laser beam in a medium (local problem).

Three-level model.

Photoionisation of an optically thin layer of Yb vapours, consisting of one given even isotope, by 3 linear polarised laser beams, coincident in space, was considered analytically and by computations.

Let the states $m = 1,2,3$ of an atom involved in photoionisation. The transitions $1\rightarrow2$, $2\rightarrow3$ are induced by laser radiation of intensities I_{12} and I_{23}. The last (third) state is photoionised to a continuum $m = i$ via an autoionising state by laser of intensity I_{3i}.

In this case, the rate equations for the level populations are

$$dN_1(v)/dt = - B_{21}(v)N_1(v)+[B_{12}(v)+A_{12}]N_2(v), \tag{2a}$$

$$dN_2(v)/dt = B_{21}(v)N_1(v) - [B_{32}(v)+A_{12}+B_{12}(v)+\Gamma_2]N_2(v)+[B_{23}(v)+A_{23}]N_3(v), \tag{2b}$$

$$dN_3(v)/dt = B_{32}(v)N_2(v) - [B_{23}(v)+A_{23}+B_{i3}(v)+\Gamma_3]N_3(v). \tag{2c}$$

Here, $A_{mm'}$ is the probability of a radiative (spontaneous) transition $m'\rightarrow m$; Γ_m is the rate of decay of a state m into states ignored in the model (open model [105-108]);

$$B_{mm'}(v) = \int_{-\infty}^{\infty} d\omega \, \sigma_{mm'}(\omega - kv) \cdot I_{mm'}S_{mm'}(\omega)/\hbar\omega, \quad (m, m' = 1,2)$$

is the rate of the transition $m'\rightarrow m$ induced by laser radiation of intensity $I_{mm'} = I_{m'm}$; $S_{mm'}(\omega)$ is the spectral function of the laser radiation inducing the transition $m'\rightarrow m$; $N_m(v)$ is the density of number of particles in the state m which have a velocity component v along the direction of the light beam; $k = \omega/c$ is the wave number ; $\sigma_{mm'}$ are the phototransition cross sections; $B_{i3}(v) = \sigma_{i3}I_{i3}/\hbar\omega_{i3}$ is the rate of photoionisation from the upper level. It is assumed that the width of the corresponding autoionising state is much greater than the spectral width of the laser radiation or the Doppler width of the $m'\rightarrow m$ transition.

The number of ions generated per unit volume and unit time is

$$dN_i/dt = \int_{-\infty}^{\infty} B_{i3}N_3(v)dv. \tag{2d}$$

The system equation (2) is valid if we can ignore the collisions altering the velocity, i.e. the transitions $v'\rightarrow v$. It is usually justified for low-density gases.

Under strong saturation conditions $B_{12}(v) \gg A_{12}$; $B_{23}(v) \gg A_{23}, B_{i3}, \gamma_3$ (the most interesting for AVLIS) the fraction of generated ions η_i is given by

$$\eta_i(t) = 1 - \exp\left\{-(\sigma_{i3}g_3/g\hbar\omega)\int_0^t I_3(t)dt\right\}.$$

Here g_m is the statistical weight of the m-th level; $g = g_1 + g_2 + g_3$ is the total statistical weight of the states involved in photoionisation. The number of produced ions is governed by the energy of laser pulse $\int_0^\infty I_3(t)dt$ and is independent of the pulse profile.

The numerical calculations based on the three-level model were done. In this model the time dependence of lasers beams intensity are assumed to be trapezoidal. The typical half-height duration of pulses was 15-20 ns, the delays between pulses varied in range 0-80 ns. The intensity of radiation of the laser in a pulse was chosen usually so that for a beam cross section area about 1 cm^2 the average power of the pulse-periodic laser operating at a repetition frequency 10 kHz ranges from 0.5 up to 50 W.

The laser spectrum in the model one could set arbitrary as well as the atom distribution on a velocity projection to the direction of a laser beam. In simulations presented here is usual assumed that all spectral functions are Gaussian.

An analysis of the time dependence of the exited states populations shows, that the radiation of different intensity are best to use for different ionisation stages. In the case of a pulsed pumping the saturation (equalisation of levels populations) occurs, when the rate of induced transitions becomes comparable with an inverse time of a laser pulse. In a case under discussion the first step is saturated at $I_{12} \approx 90$ W/cm^2; second - at $I_{23} \approx 4000$ W/cm^2. The efficiency of ionisation can exceed 50 % if the intensities are $I_{12} \approx 0.5$ kW/cm^2; $I_{23} \approx 20$ kW/cm^2; $I_{3i} \approx 100$ kW/cm^2.

Certainly these outcomes should be reconsidered for the optically thick transitions and for the medium in a cavity (see 3.2).

Multilevel multi-isotope model.

A multilevel model was advanced to determine a photocurrent depending on the wavelengths of the lasers, stimulating the first and second transitions. This model takes into account simultaneous ionisation of the all isotopes, corresponding to the spectrum of used lasers and it, accordingly, enables to determine the enrichment resulting from the selective photoionisation.

A set of equations like (2) was used for different isotopes and different velocity groups. The natural isotopic composition was assumed. The Yb vapour was characterised by the density and the atom distribution on a velocity projection along the direction of the laser beams propagation. The atoms in the medium were divided into $n = 5 \div 15$ groups in accordance with the velocity projection to a direction of a laser beam. For each velocity group the balance equations were used. The $50n$ equations were solved. The velocity distribution of atoms could be set arbitrary. The results of computations shows that in order to ensure the high selectivity of the ^{168}Yb ionisation the second step laser wavelength should lies within the range 581.0785 ± 0.0005 nm.

For various reasons the laser radiation spectrum can have wide wings. Since the initial abundance of the target isotope is very low, the radiation represented by these wings may have a significant influence of the ionisation selectivity. For these reasons the model of a multistage selective photoionisation of multilevel ^{168}Yb atoms [8] was extended in Refs. [4, 5, 10]. The presence of wide wings was taken into account by modelling the profile of a laser line so that it has a 'pedestal' specified on the basis of additional considerations.

The results of these calculations (Fig. 3) show that if the wing of a laser spectrum extends in the direction of long wavelengths by 0.007 nm (6 GHz), then its influence on the isotope composition becomes significant even when the spectral energy contrast is ~ 100. This contrast is understood to be the ratio of the spectral density of the energy emitted by a laser at the centre of a line to the maximum spectral density of the energy emitted by the same laser in the wings of the line when the detuning is $|\Delta v > 1.5$ GHz. The 1.5 GHz frequency corresponds to the detuning of the energy level of the nearest isotope in the second stage of the excitation process.

J (rel. units)

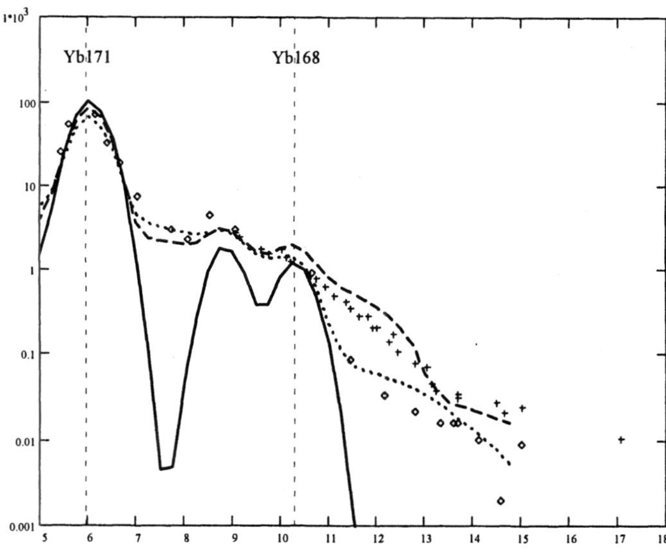

Δv/GHz

Figure 3: A photocurrent (relative units) via laser frequency detuning Δv of the frequency of the second (selective) $4f^{14}6s6p\ ^3P_1 \rightarrow 4f^{13}6s^26p$ (7/2,3/2)$_2$ transition. Δv is measured from frequency of this transition in ^{174}Yb. Solid curve corresponds to calculations in absence of 'pedestal' in laser radiation spectra; diamonds and crosses correspond to series of experiments at different conditions. The dashed curve corresponds to a 'pedestal', at which the experimental data (crosses) agree satisfactorily with calculations ('pedestal' has a relative magnitude 0.02 at $\Delta v = \pm 5$ GHz and 0.05 at $\Delta v = 0$). The dotted curve corresponds to a 'pedestal', at which the experimental data (diamonds) agree satisfactorily with calculations ('pedestal' has a relative magnitude 0.002 at $\Delta v = \pm 7$ GHz and 0.02 at $\Delta v = 0$). According to the calculations at the peak of the current corresponding to the Yb168, its concentration is 51 % for the dashed curve and 54% for the dotted curve

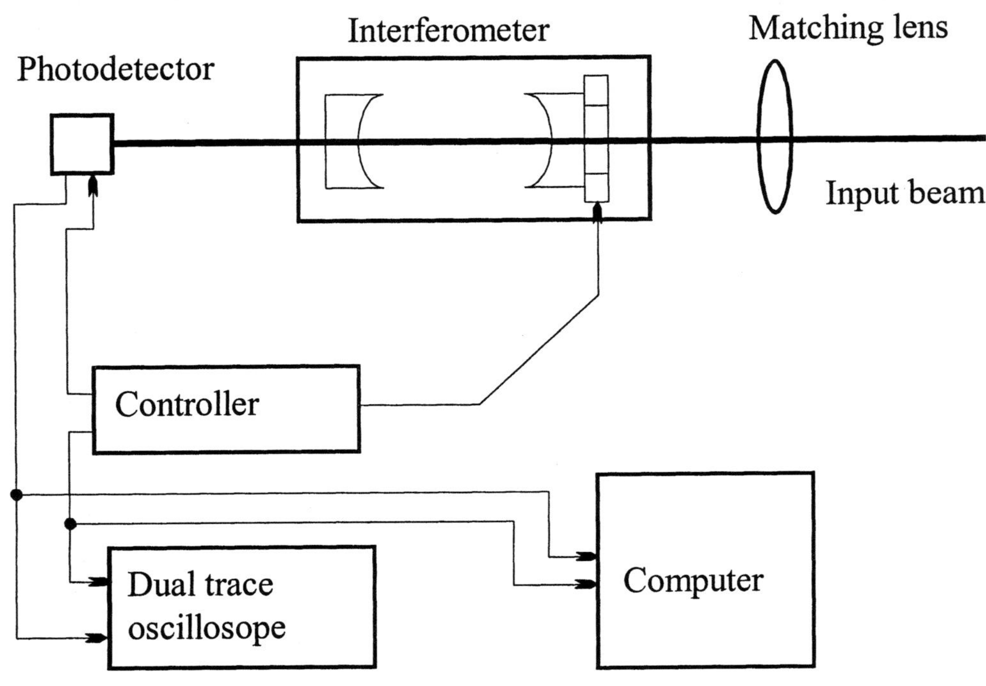

Figure 4: Computer analyser of the laser radiation spectrum

2.3. The dependence of the photocurrent on the spectrum of the selecting laser transition

In view of the small contents of ^{168}Yb in a natural mix it is necessary to emphasize the importance of the spectral contrast of the lasers radiation of the second stage of excitation. The intensity of radiation in a wing ($\Delta v \sim 6$ GHz in short-wave area) of the second laser line should be weaker than intensity of the line centre in a few hundreds times. Otherwise an essential ionisation of the 171 and 173 isotopes takes place.

So, we have measured a dependence of photocurrent on the various laser wavelengths [1, 3-5, 10]. The dependencies of the photocurrent on the wavelengths of various lasers were determined [4, 5, 10] (the facilities used were described in Refs. [1,3], and will be considered in Section 6). A comparison of the experimental results with calculations showed (Fig. 3) that wide wings might play a significant role. Consequently, measures were taken to improve the quality of the laser spectrum and to monitor the positions and widths of the laser lines. Moreover, a mass-spectrometric monitor was used in the course of isotope production.

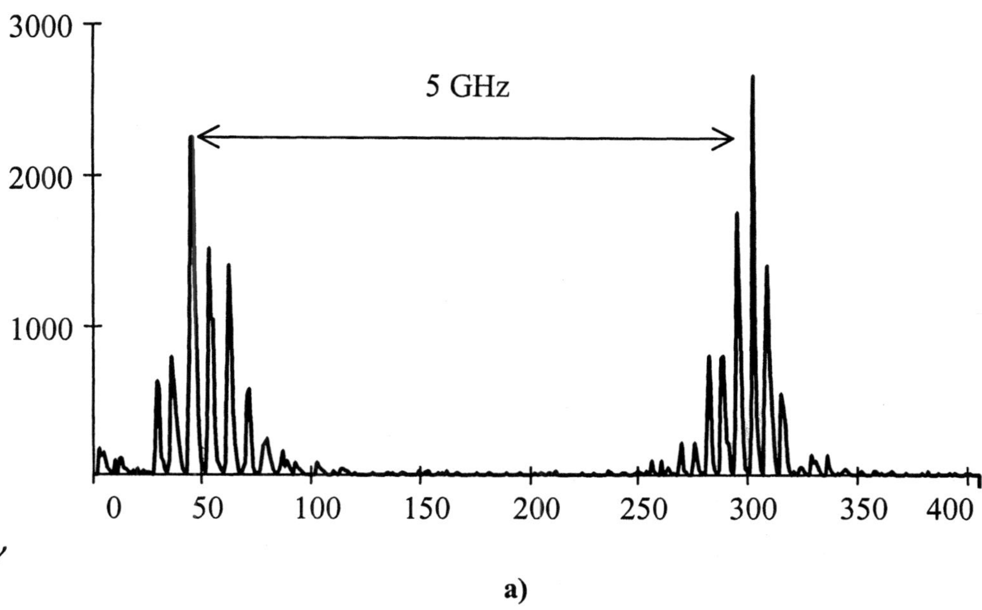

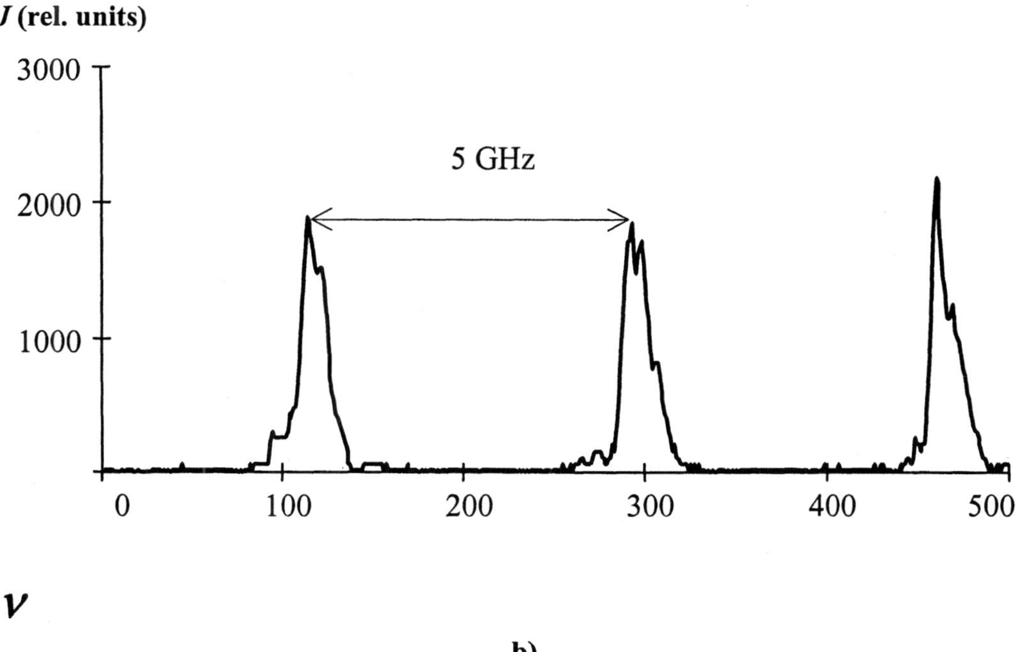

Figure 5: Emission spectra of the laser used in the second excitation stage, recorded in the absence of an integrating *RC*-chain (a) and in the presence of this chain (b). *N* is the channel number

The wavelengths of the three dye lasers employed were measured and monitored with a precision wave meter, developed at the Institute of General Physics of the Russian Academy of Sciences, which ensured that the error in the wavelength measurements was just $\Delta\lambda/\lambda \approx 10^{-7}$. This wave meter, constructed from four Fizeau interferometers with different bases, made it possible to estimate roughly the spectrum of a laser and, in particular, the half-width of the line profile, since the instrumental function of the most accurate of the interferometers had a half-width ~ 400 MHz and a free spectral range $\sim$ 1.7 cm^{-1}. However, the finesse of this instrument was fairly low (it could be used to measure the spectral contrast up to 12). Moreover, the line profile obtained with the aid of this instrument represented integration of ~ 100 laser pulses, so that it was not possible to draw conclusions about the signal/noise ratio and the true structure of a laser emission line on the basis of such measurements.

The use of laser radiation with a high (in excess of 200) spectral energy contrast made it possible to reach a concentration of ^{168}Yb ions in a plasma in excess of 90% when the natural abundance of this isotope in the original material was 0.14% (see Section 6).

3 Laser beams arrangement

3.1. Distance travelled by laser beam in a medium

The distance at which the intensity of the radiation I_0 entering the medium falls, under steady-state conditions, to approximately the saturation value $I_S = \hbar\omega A/\sigma^{ph}$, can be estimated by the formula $l \approx I_0/I_S\kappa_0$. Here, $\kappa_0 = \sigma^{ph}N$ is the unsaturated absorption coefficient, σ^{ph} is the resonant phototransition cross section, N is the density of absorbing atoms. However, such estimation is too optimistic. We must not use the peak intensity but the total energy of laser pulse, because pulse is usually short.

We tackled this problem by solving simplified set of equations, describing the propagation of the resonant radiation $I(z, t)$ on an axis z in a medium of two-level atoms:

$$\frac{1}{c}\frac{\partial I}{\partial t} + \frac{\partial I}{\partial z} = \sigma^{ph}I(N - 2N_1), \qquad \frac{\partial}{\partial z}(N - 2N_1) = (N - 2N_1)(2\sigma^{ph}I/\hbar\omega),$$

where N_1, $N_2 = N - N_1$ is the density of the active atoms in the lower 1 and upper 2 states. Computations show, that under the strong saturation condition $I_0 \gg I_s$ (this case is of greatest interest) the results look as follows. A steep leading edge of width $\Delta z_F \approx 1/\kappa_0$ $\approx 1/\sigma^{ph}(N_1 + N_2)$ is formed. This front travels with the velocity $v_F = c/(1 + I_E/I_0)$, which can be deduced by energy conservation reason. The quantity $I_E = \hbar\omega cN/2$ occurs here does not coincide with saturation intensity I_S. At the most interesting conditions I_E is much greater I_S ($I_E/I_S = c\sigma^{ph} N/2A_2 \sim 10^4$, if one assume $N_1+N_2 = 3\cdot10^{12}$ cm^{-3}, $\sigma^{ph} \approx 3\cdot10^{-13}$ cm^2).

At the same time, the trailing edge of the pulse continues to move with velocity of light c. The tail catches up with front during $\tau_I \equiv \tau v_F / c$ and over distance $z_E = \tau v_F = \tau_I \cdot (2I_0 / \hbar\omega N)$ (here τ is the laser pulse duration) the pulse vanishes. If one assume $\tau \approx 15$ ns, $I_0 \approx 1.7\ 10^4$ W/cm^2 (average power 1 W per 1 cm^2 at repetition frequency 10 kHz) the travelled distance would be $z_E \approx 2$ m.

For $\tau >> z_E / v_F$ the estimation given above for the steady-state case can be used. The computations for various intensities are in a good agreement with the approximate formulas given above.

3.2. Radiation in 0D-cavity

Two-level absorbing medium in a cavity.

Efficient utilisation of laser radiation is seriously hindered by the difference between the photoexcitation and photoionisation cross sections of atoms, which differ in several orders of magnitude. At a distance corresponding to total absorption of laser radiation in the first stage (when atoms are excited from the ground state) only a small proportion of the final-stage radiation (which ionises the atoms) is absorbed. The main method for solving this problem involves the use of a cavity as a multipass system, which delays that radiation which is absorbed weakly by the medium employed. The efficiency of utilisation of a cavity in multistage ionisation of the ytterbium was demonstrated in Ref. [9] by considering the example of a zero-dimensional model.

Let us assume that a diverging laser beam enters through a small aperture in one of the cavity mirrors and then experiences multiple reflections, making several passes through the cavity and thus filling a certain volume (Fig. 6). The number of such passes is limited by the absorption in the medium, by the reflection losses, and by the escape of the radiation from the cavity either beyond the mirror edges or back through the entry aperture.

In the 0D approximation the problem is considered as effectively homogeneous. The cavity characterised by a damping decrement, and the radiation in it represented by its total energy. For convenience of comparison with a real situation we shall describe the radiation by intensity I averaged over the effective volume occupied by this radiation. We assume that it is the homogeneous an extended volume of length L and cross-sectional area S. We shall express the damping decrement $\gamma_n = \gamma/n$ ($\gamma = c/L$) in terms of the effective number of passes n in the empty cavity.

We shall consider situation when the radiation is injected into a cavity during time much smaller the damping time of cavity n/γ. The absorption is characterised by unsaturated absorption factor κ_0 and intensity of saturation I_S. For intensity of radiation in 0D approach in a dimensionless form we have the equation:

$$\frac{di}{dt} = -i \cdot \left(1 + \frac{\eta n}{1+i}\right), \quad i(0) = i_0. \tag{3}$$

Here $i = I/I_S$ is intensity in terms of saturation intensity; $\tau = \gamma_n \cdot t$ is the time in terms of the radiation life time in the resonator; $\eta = \kappa_0 \cdot L$ is the optical thickness for one pass; $i_0 = I_0/I_S$ is the dimensionless intensity of the injected beam. The initial intensity I_0 is connected with the surface density $\Delta E/S$ of the energy ΔE injected into the cavity by the ratio: $I_0 = \Delta E \gamma / S$.

The solution of this equation is:

$$\tau(i, i_0, \eta n) = \frac{1}{1 + \eta n}\left[\ln\left(\frac{i_0}{i}\right) + \eta n \ln\left(\frac{1 + i_0 + \eta n}{1 + i + \eta n}\right)\right].$$

The energy of radiation, absorbed by the resonator, is determined by expression:

$$E_R = \frac{I_{sat}}{\gamma}\int_0^{i_0} \tau(i, i_0, \eta_n)\,di = \Delta E(1 - \varepsilon),$$

where

$$\varepsilon = \frac{\eta n}{i_0}\ln\left(\frac{1 + i_0 + \eta n}{1 + \eta n}\right) \tag{4}$$

is the fraction of radiation absorbed by the medium (Fig. 7).

We shall assume parameters typical for the system discussed below: lasers with average power $P_m \sim 1$ W with the pulses repetition frequency $f \approx 10$ kHz and duration of a pulse $\Delta t \approx 15$ ns. It corresponds to energy $\Delta E \approx 10^{-4}$ J injected by one pulse. We shall also assume, that a diameter of the radiation beam is $d \approx 1.5$ cm ($S \approx 1.8$ cm^2), and the cavity length is $L \approx 1$ m.

Let's assume the vapour density in a selective ionisation zone equal to 10^{13} cm^{-3}. Accordingly, the sum of the isotopes concentration absorbing at the first transition will be $N = [(2 \cdot 0.16/3) + 0.14] \cdot 10^{13}$ cm$^{-3} \approx 2.5 \; 10^{12}$ cm^{-3}. Thus the typical values of the quantities occurring in expression (3) are: $i_0 \approx 5.7 \cdot 10^3$; $\gamma_n \approx 6$ MHz (for $n = 50$); $\eta \approx 37$ for $1 \rightarrow 2$ transition, $\eta \approx 0.8$ for $2 \rightarrow 3$ transition.

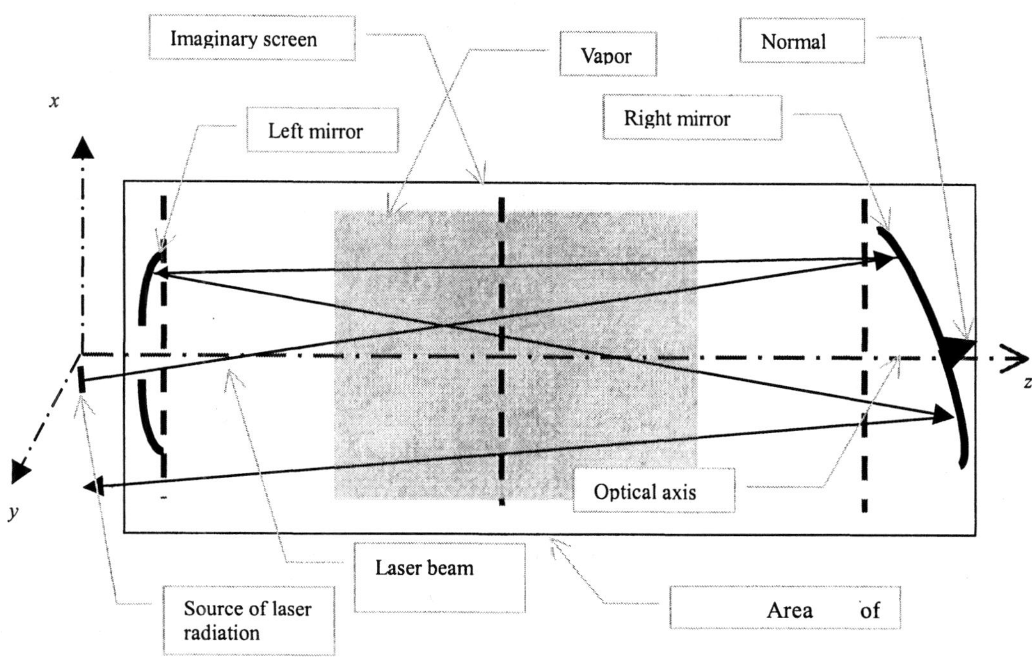

Figure: 6. Scheme of 3D cavity and a laser beam trajectory

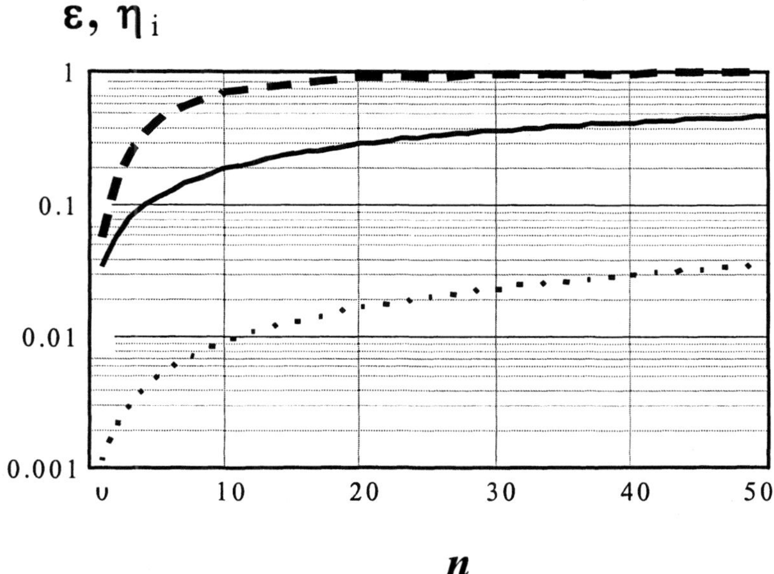

Figure 7: Fraction ε of radiation absorbed by the medium and efficiency η_i of the [168]Yb ionisation (dashed line) predicted by the 0D multilevel model of selective ionisation via effective number of passes n through the empty cavity under the same conditions. Solid line is for the first stage laser, $\eta \approx 37$; dotted line is for the second stage laser, $\eta \approx 0.8$; ε is given by formula (4)

Multilevel model of selective photoionisation of ytterbium in 0D cavity.

The equation (3) is valid only in quasi-cw approximation, when it is possible to neglect the time derivatives from level populations. Moreover, in considering the selective multistage ionisation we cannot limit ourselves to the two-level approximation. Therefore, we have constructed a model of selective photoionisation of ytterbium, similar to that discussed above. This model is distinguished from above one by the fact that the lasers radiation intensities at various transitions are not considered as given functions of time, but are calculated taking account of the absorption in the medium and its degradation in the cavity.

In the equations for the intensities we replace the expression $\alpha \kappa_0 \left(1 + I / I_{sat} \right)$, which takes into account a light absorption by the medium in the quasi-cw approximation, directly with the expressions proportional to populations of the relevant levels:

$$\frac{dI_{m,m'}}{dt} = -I_{m,m'} \cdot \left(\gamma_n + c\sigma^{ph}_{m,m'} N_{m'} - c\sigma^{ph}_{m',m} N_{m'} \right).$$

Here N_m, $N_{m'}$ are the populations of states m, m'; $\sigma^{ph}_{m',m}$, $\sigma^{ph}_{m,m'}$ are the cross sections of phototransition $m \rightarrow m'$ and $m' \rightarrow m$ respectively.

It is evident from Fig. 7 that the fraction η_i of the ^{168}Yb atoms ionised per pulse depends on the number of the cavity passes in almost the same way as the fraction ε of the energy absorbed in the second transition. This is to be expected, since the power of the ionising laser is selected to be sufficiently high so that the rates of the $2 \rightarrow 3$ and $3 \rightarrow i$ transitions are comparable. Such an analysis was carried out in the course of first stage of the design of the installation described below.

3.3. Radiation in 3D-cavity

3D models

The 0D approximation adopted for the radiation in the cavity makes it possible to demonstrate quite clearly the advantage of a multipass cavity and to obtain (to within a factor of 2) the volume-average values of the quantitative characteristics of the medium and of the radiation as a function of the number of passes through the empty cavity n. However, in a real situation the actual number of passes through the empty cavity is not a function of the mirror absorption alone. 3D models are needed in order to propose recommendations on selection of a specific cavity configuration.

A general theory of cavities is quite well developed (see, for example, Ref. [109]), but problems resulting from the use of cavity in the AVLIS method are highly specific. The main interest lies in calculation of the overlap of laser beams in the ionisation zone and the kinetics of ionisation with the resultant spatial distribution of its density. This determines the main parameters, which are of interest in the AVLIS method: the ionisation efficiency and the number of ions produced by one pulse.

Three 3D models were developed [11]. The first model corresponds to the empty cavity. Most coarse information we get about a laser radiation in empty cavity is: the number of passes n; the area of beams overlapping, and distribution of the energy, which was passed in both directions through various planes perpendicularly to the optical axis. And so, some information was presented on imaginary screens situated perpendicularly to the z-axis (see Fig.6). In particular, we display the intensity distribution on a mirror surface. This simulation is useful for preliminary laser alignment ensuring spatial overlapping of laser beams.

Within the framework of the second model spatial-temporary ionisation of ytterbium vapours was considered on the basis of multilevel model discussed above. The absorption of light by a medium, however, was not taken into account in this model. Such statement of a problem is of interest for the case of rather small vapour density and not big length of an absorbing medium. It allows also to get a limiting current on a collector, which can be obtained, for the given geometry of cavity and laser beams and for the given vapour density.

The method of 'large particles' (photons) developed specially for the problem of generation of radiation of laser oscillators and amplifiers [110-112] was employed in these models. Each of particles was characterised by the velocity direction, spectral frequency, spatial co-ordinates and fictitious weight. The photons were moved at each time step. After calculation of intensity in the appropriate spatial grid the set of the balance equations for levels populations was solved. After that the photons were moved again.

Outcomes of simulations are the spatial-temporal intensity of lasers light and amount of ions of each isotope on a space grid. On the base of these data the limit ionic current, the plasma isotopic composition and some other characteristics were calculated. The basic resulting characteristics are the number of ionised atoms and the effectiveness of the target isotope ionisation.

The third model differs from the second one by that the absorption of a radiation at the first stage of ionisation is taken into account. The set of the balance equations for level populations together with the radiation intensity equations was solved at each time step. The particle weights were modified according to the obtained solution. An essential moment takes place if one solve the kinetics equations together with the equation for laser light intensity (for given space point, for given group of frequency and for given group of velocities of atoms). In this case kinetic equations for different isotopes are not uncoupled. A common set of equations (51 equations) for each velocity group therefore must be solved.

Third model gives us an actual current at the collector because we know the ions extraction factor by plasma simulations (see below). The simulations give the opportunity to predict the increasing of ionic current with growth of absorbing medium length.

Simulations and experimental data.

Outcomes one of typical simulations illustrates Fig. 8 (the initial data are presented in a brief form). In the example presented here almost flat mirrors were assumed to be tilted relative one another at small angle. The light beams directed under a small angle to optical axes, being reflected from mirrors, have a diffusion-like travel along an axis x, directed along the atom flux through zone of ionisation. After some number of reflections the motion changes the direction. In area of this turn-round the effectiveness of ionisation is maximum (Fig. 8c). Through, laser radiation pulses are passing twice x points, located before area of a turn-round (see Fig. 8b).

The dependence of an average ionic current via an average power of the laser radiation injected in a resonator illustrates a Fig. 9. Even the 0D model gives not bad results for the second and third stages of ionisation. The small falls of experimental values of currents in a fig. 9b are connected, apparently, with small disadjustment of the laser beams. Attenuating plates were placed on a drum. The preliminary adjustment was produced for of maximum laser power. It could be, for example, due to non strict parallelism of plates absorbing a laser radiation.

When the absorption of a radiation in a medium is essential, the 0D-model gives too sharp passage from weak absorption to saturation. Besides, it is necessary to remember that in the 0D model some experimental data (number of passes of a radiation in the empty cavity n and an effective cross-section of the area of the laser beams overlapping) are used. In 3D models these magnitudes follow from simulations.

Calculated value of a maximum current for all models is a little bit above experimental one (see subscript to a Fig. 9). These simulations do not take into account that fact, that not all ions was extracted from plasma. From comparison of a maximum value of a current obtained in these experiments with 3D-model absorption simulations follows, that extraction factor of ions from plasma was 65%. This value is in accord with simulations of extractor properties.

4 Flow of vapours

Reliable modelling of the AVLIS technology requires consideration of the properties of a vapour source, which are needed to determine the vapour density and the distribution of the velocities of atoms in the ionisation zone. Although studies of the formation of atomic beams have a long history (see, for example, Refs. [113, 114]), some aspects important for the AVLIS technology have been almost completely ignored.

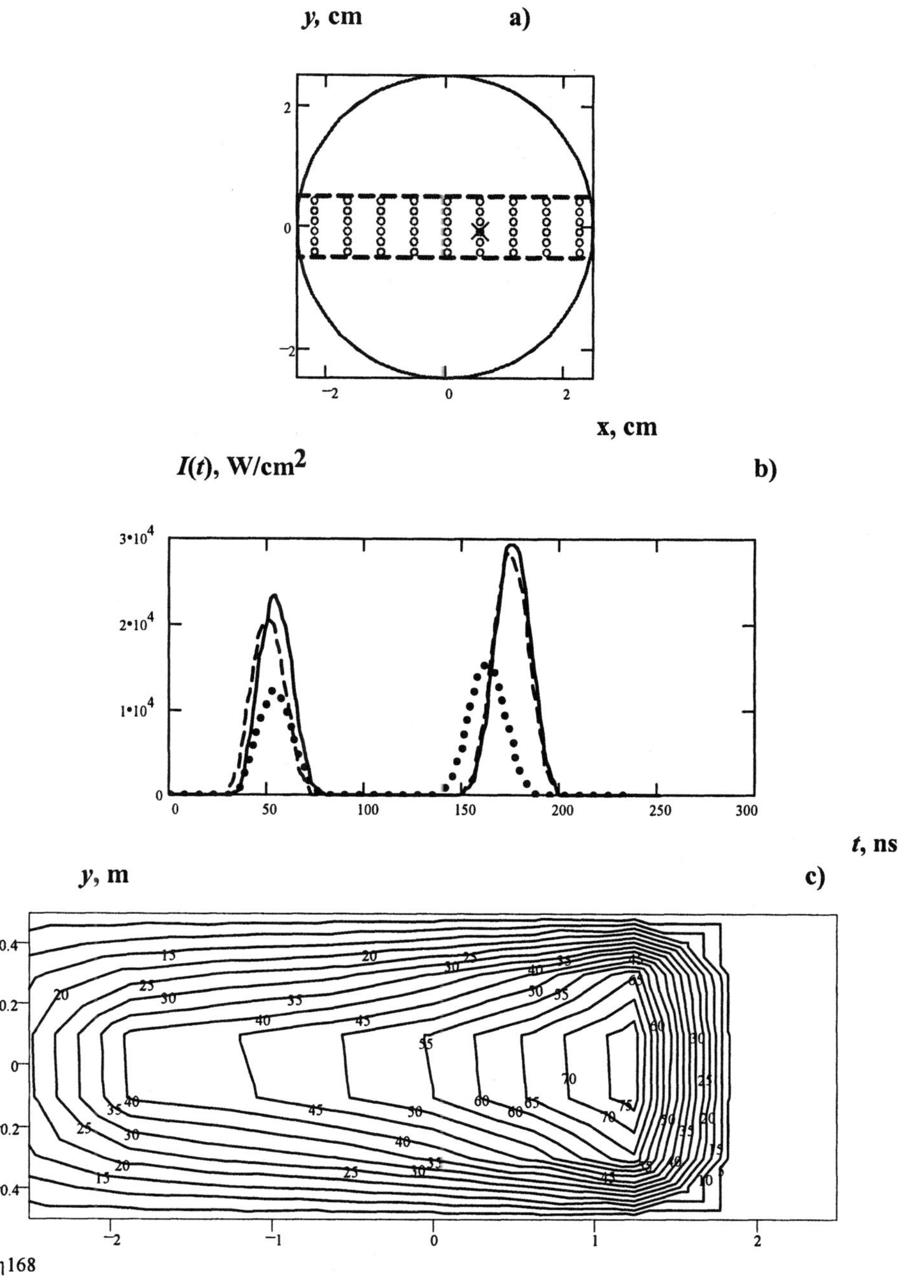

Figure 8: Example of 3D simulations for a filled cavity (ignoring absorption):

a) geometry of the absorbing medium and the reflecting surface of the mirrors (the circles mark points of a grid);

b) time dependence of the radiation intensity at the right mirror in the point marked by an oblique cross at the Fig. 8a. Dotted curve represents the first laser radiation intensity, dashed curve represents the second laser radiation intensity, solid curve represents the third laser radiation intensity;

c) the contour lines of the effectiveness of ionisation ^{168}Yb at the cross-section of the cavity (in percentage).

Some initial data of the run presented at Fig. 8.
Cavity characteristics:
mirror diameter   5 cm;
diameter of input hole   1 mm;
vectors along mirrors normal $s_1 = \{-0.39,0,200\}$, $s_2 = \{-0.36,0,200\}$.
Lasers beams characteristics:
Laser 1:
the centre of the source of the laser radiation: -2.0, 0,   0;
diameter of the source of the laser radiation and of a spot at the right mirror: 0.5, 0.51;
projection of the vector which give the initial beam direction: 0.07, 0, 200.
Laser 2:
the centre of the source of the laser radiation: -1.8, 0, 0.1;
diameter of the source of the laser radiation and of a spot at the right mirror: 0.5, 0.51;
projections of the vector which give the initial beam direction: 0.04, 0, 200.
Laser 3:
the centre of the source of the laser radiation: -1.8, 0, 0;
diameter of the source of the laser radiation and of a spot at the right mirror: 0.5, 0.51;
projections of the vector which give the initial beam direction: 0.06, 0, 200.
Lasers pulses characteristics:
repetition frequency 10^4 Hz;
delay, front, constant top and collapse of trapezoid impulse (ns)   0, 10, 10, 10.
An average power of lasers inputted in the cavity: $P_1 = 1.5$ W; $P_2 = 1.3$ W; $P_1 = 3.4$ W.
Characteristics of absorbing media:
the length along optical axis   1 m;
vapour density $N = 10^{12}$ cm^{-3};
spectral lines width at half height   650 MHz;
vapour flux flow   2.7 10^4 cm/s.
Lasers spectra characteristics:
Gaussian-type;
spectral lines width at half   430 MHz;
3 points of spectra detuning for 'pedestal' set   -14000, 0, 14000 MHz;
relative spectra amplitude of 'pedestal'   0.01 0.1 0.01;
spectra detuning of centre of first laser line   -10293.1 MHz;
spectra detuning of centre of second laser line   3636 MHz

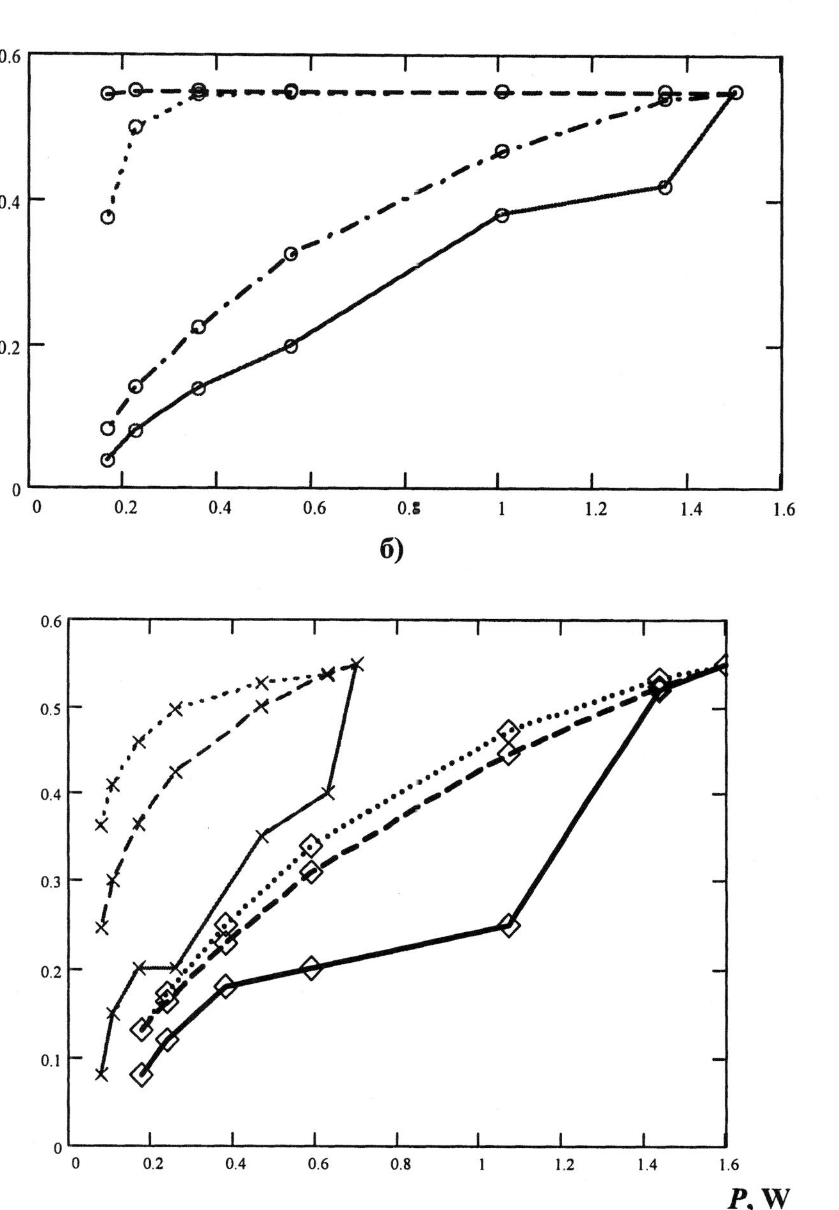

Figure 9: Comparison of simulations on various models with experimental data. Ionic current (mÅ) via average power (W) of lasers radiation injected into the cavity. Dotted curves represent 0D model; dashed curves represent 3D model ignoring absorption; dash-dotted curves represent 3D model with absorption; solid curves represent experimental data. Circles represent the dependence on the laser power of the first stage of excitation (Fig. 9b); crosses represent the dependence on the laser power of the second (selective) stage of excitation; diamonds represent dependence on the laser power of the third (ionising) stage. Other parameters are the same, as for Fig. 8.

The calculated curves are normalised so that they coincide with experimental data for maximum values of laser radiation power (0.55 mA for P_1 = 1.5 W/cm^2, P_2 = 0.7 W/cm^2, P_3 = 1.6 W/cm^2). The calculated maximum values of currents for various models are follows: 0.94 mA for 0D model; 1.2 mA for 3D model ignoring absorption; 0.84 mA for 3D model with absorption

For a number of reasons the vapour density in the ionisation zone should be low, so that the mean free path in this zone is considerably greater than the transverse size of the atomic beam. This makes it possible to consider the flow of a vapour in the ionisation zone as free-molecular and the escape of atoms to the geometric shadow region can be taken into account within the framework of the single-collision approximation [12-15]. However, the free-molecular flow approximation may fail in the region of flow of the vapour from a thermal bath. Even the criteria for departure from free-molecular flow are not clear a priori. The flow of atoms under conditions when the hydrodynamic approach (the approximation of a continuous medium) is still invalid, but the motion of atoms can no longer be regarded as free-molecular, is best modelled by a statistical box-in-a-cell method [115]. This method had been used to consider problems of the formation of flow of atoms through a slit and the escape of atoms into a vacuum [16, 17].

4.1. Free molecular flow

Formulation of the problem.

In this case we consider free molecular propagation along the channel, i.e. atoms collisions only with tube walls are accepted into attention. The atom-wall scattering was considered as diffusion one [116-118]. The consideration of an atoms flow using the assumption of a free molecular motion is limited by rather low density: $N << 1/\sigma l$, where σ is the cross section of an elastic atom-atom collision, l is a tube length (see below).

Specifically, we simulated the following situation. We assumed that the emitting surface coincides with the bottom of the tube. The velocity distribution of particles ejected from the bottom was taken in the form

$$f_T(v) = Cv_x \, exp\!\left(-M_{Yb}(v-V)^2/2T\right)$$
$$\int dv f_T(v) = 1, \quad -\infty < v_y, v_z < \infty, \quad 0 < v_x < \infty. \tag{5}$$

Here, the x-axis is directed along the axis of the tube, v_x v_y v_z are the projections of an atom velocity $\mathbf{v}$ to the respective axes, M_{Yb} is the mass of Yb atom, T is the vapour temperature in the bath, $\mathbf{V}$ is the mean velocity of the flow, and C is the normalisation constant. In case $\mathbf{V} = 0$, distribution (5) corresponds to the Maxwelian distribution function in the volume that serves as a source of particles. A similar distribution (with $\mathbf{V} = 0$ and $T_w = T$, where T_w is the temperature of the walls) was chosen for diffuse-reflected particles.

Flow from one tube.

The fraction k of particles transmitted through the tube decreases with the relative tube length l/d is in accord simple formula: $k = (1 + 0.17\sqrt{x} + 0.75x)^{-1}$, $x \equiv l/d$; here d is a tube diameter.

As it is well known, the directivity of a certain fraction of the atomic flow increases with a growth in the ratio l/d. However, the probability $J(\rho) = \int_0^\rho 2\pi \cdot j(u) \cdot u \cdot du$ of atoms hit in a circle of radius $r = \rho x_0$, at $r \sim x_0$ remains practically the same even for considerable differences in l/d. The fact is that only a small amount of particles are contained in the small angle area. In particular, angle $\theta_{0.5}$ that include a half of the atomic flow transmitted through the tube, only slightly depends on a relative tube length l/d (see Fig. 10). At the same time the angle, corresponding to half-height of distribution of the transmitted flow considerably depends on a relative tube length l/d.

The fact is important, that the greater fraction of particles is scattered to large angles during the passage through the long tubes. We had to take it into account in a design of an ion extraction system. The particles, scattered by large angles can contaminate the ions collector.

A flow from a set of tubes.

Usually the distribution of a total flow from a set of tubes, filling an extended rectangular $\Delta L \times L$ ($\Delta L \ll L$), is of interest. Here we shall be limited by consideration of two cases: a) of one tube line and b) of 20 tube lines.

In both cases the length of tube set was $L = 1$ m, and the tube length was $l = 10$ mm. In the first case one line of 500 tubes with a diameter $d = 2$ mm ($\Delta L = 2$ mm) was considered. In the second case 2000×20 tubes with the diameter $d = 0.5$ mm arranged within $\Delta L = 15$ mm was considered. The system of co-ordinates is placed in the centre of the rectangular, formed by tubes; the axis x is directed parallel to the tubes axes, the axis y is directed transverse the rectangular, and the axis z is directed along the rectangular.

As well as it was necessary to expect the flow density distribution on a z-axis, looks like a rectangular one. The distribution of the flow density in this case is almost the same, which is given by expression

$$f(z) = \frac{2x_0^3}{\pi L} \int_{-L/2}^{L/2} \frac{du}{\left[(u-z)^2 + x_0^2\right]^2}$$

corresponding to an infinitely narrow and infinitely thin slit ($\Delta L \to 0$, $l \to 0$). (The explicit expression for the integral in the above formula is very cumbersome, and we do not present it here.)

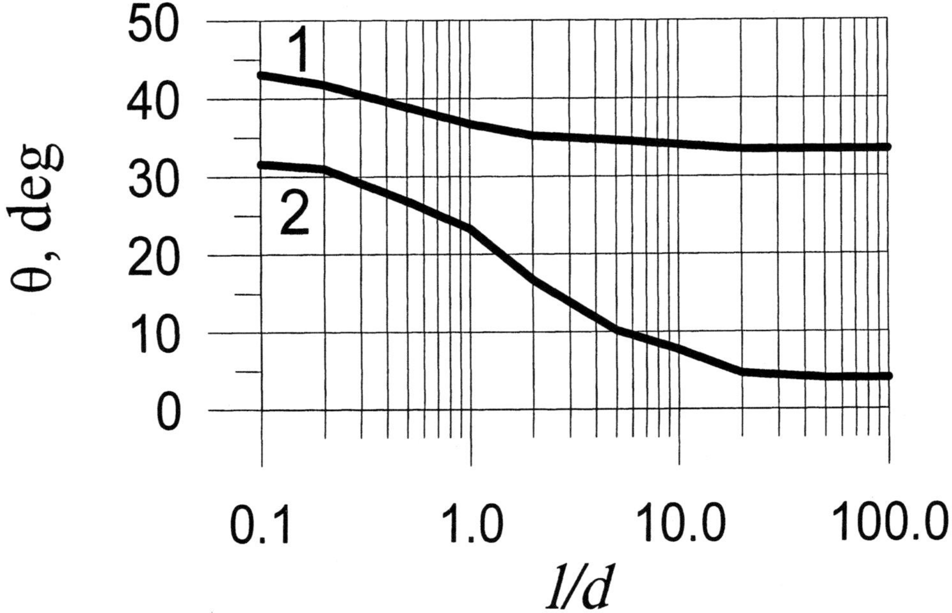

Figure 10: Angles θ characterising a particles flow emerging from the tube via a relative tube length: 1 - angle corresponding to the half of the total flow; 2 - angle corresponding to the half-height of the flow distribution.

In case of 20 tube lines the distribution on an y-axis is a few more homogeneous in comparison with distribution from one tube line (see curves 1,2 in Fig. 11). However the fraction of atoms, scattered to a cross direction because of the effects mentioned above is not small, so the complete probability $P(y)$ of atom hit, transmitted through a strip of y width, practically coincides for one tube line and many tube lines (see curves 3,4 on Fig. 11). For comparison, the following distributions are given:

$$j(y) = \frac{1}{2x_0} \frac{1}{\left[1 + (y/x_0)^2\right]^{3/2}},$$

$$P(y) = \frac{1}{2} \frac{y}{\sqrt{y^2 + x_0^2}}.$$

They correspond to a source as an infinitely narrow, infinitely thin and infinitely long split ($\Delta L \rightarrow 0$, $l \rightarrow 0$, $L \rightarrow \infty$, and curves 5 and 6 in Fig. 11).

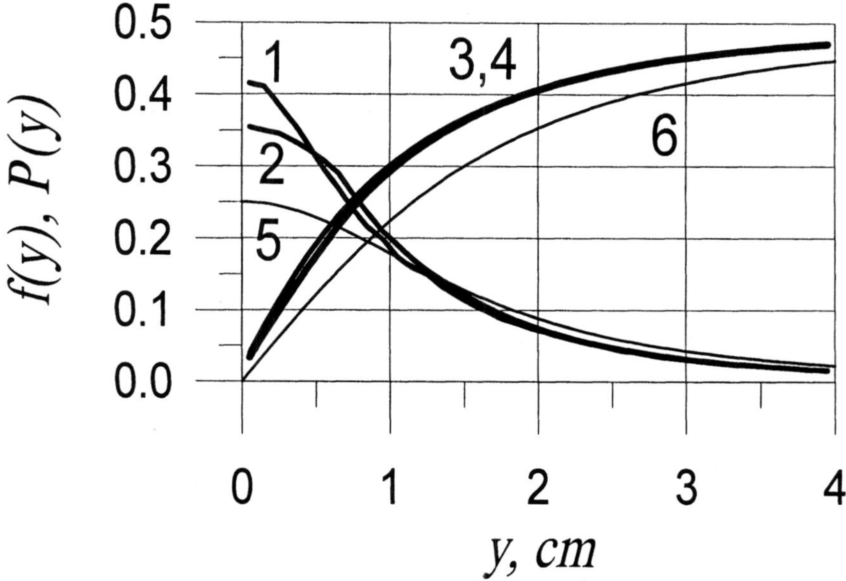

Figure 11: Distribution of an atoms flow transmitted through the set of parallel cylindrical tubes. The set of tubes forms a rectangular with width ΔL and length L.

The curves *1,2* and *5* display the probability density $f(y)$ that atom falls within unit area in the plane perpendicular to the tubes axes located at the distance $x_0 = 20$ mm from tubes outlet as a function of a distance from the plane of symmetry (along the *y*-axis) in the cross section corresponding to the half width of the tubes set (in transverse direction). Curves *3,4* and *6* display the probability $P(y)$ that an atom falls within a band with width of *y*. Curves *1,3* corresponds to one line of tubes; curves *2,4* corresponds to 20 lines of tubes and curves *5,6* corresponds to the equilibrium emission of atoms from an infinitely narrow, infinitely thin and infinitely long slit.

4.2 Flow of vapour in the ionisation zone

Calculations ignoring collisions.

The vapour density in the ionisation zone should be low, so that the mean free path is considerably greater than the transverse size of the flowing vapour in order to avoid collector contamination. In view of this, the distributions of the densities of the atoms and of the vapour in the ionisation zone can be assumed to be the same as in the case of free-molecular flow. By analogy with the well-known conservation of the radiation intensity along an optical ray in a transparent medium, in the case of free-molecular flow we can assume that $f(\xi,\mathbf{v}) = f(\xi + \mathbf{v}\tau,\mathbf{v})$. Here, $f(\xi,\mathbf{v})$ is the distribution of atoms in terms of the velocity vector $\mathbf{v}$ at a point in space ξ, τ is an arbitrary time interval. Consequently, we can find the distribution function $f(\xi,\mathbf{v})$ at an arbitrary point of the ionisation zone and consider it as the point of intersection of the rays passing through the surface emitting atoms. Integration over the velocities gives the vapour density.

Single collisions.

The consideration of single collisions is necessary in order to account a vapour flow into an area of the geometrical shadow, which is provided usually by the system of diaphragms [13,14].

For the scattered atom flow from homogeneous flux of width D infinite along axes y, z the simple expression was obtained:

$$F(x) = \sigma\sqrt{\frac{2T}{M_{Yb}}}DN^2\chi(x/a), \quad \chi(u) = 0.018\,u^4\,/\left(1+1.62u^2+0.1u^4\right), \text{ at } 0<u<3. \qquad (6)$$

It was considered, that the atoms fly in parallel to x axis and that the velocity distribution was "semi-Maxwell" one:

$$f_x(v) = 2(M_{Yb}/\pi T)^{1/2}\exp\!\left(-M_{Yb}v^2/2T\right)\ \ -\infty < v_y, v_z < \infty,\ 0 < v_x < \infty.$$

Here $F(x)$ is the scattered flow density at line, which is parallel to the axis of vapour flow, and placed at a distance a from the diaphragm edge; x is a distance above diaphragm plane.

Using a model of a collinear beam of atoms with a semi-Maxwell velocity distribution we can estimate the relationship between the relevant quantities. However, quantitative predictions of this model should be verified with computations. The fact is that under the real conditions the atoms beam has a noticeable divergence. It considerably increases the number of scattered atoms within the area of geometrical shadow.

We performed simulations for the scattered flow assuming that the source has the shape of a plane slit with a width of 2 cm. The diaphragm width was $D = 1$ cm (see Fig. 2), slit-diaphragm distance was equal 2 cm. We present the flows for two values of a ($a = 1.5$ cm and $a = 3.5$ cm). The obtained distributions were represented in the form (6) and functions $\chi(x/a)$ were obtained (Fig. 12). The results of simulation are presented only for the area of geometric shadow of the forward flow.

As one might expect, for a non-collinear atomic beam, the scattered flow cannot be characterised by universal function $\chi(x/a)$. The different χ functions correspond to various a. In addition, the number of particles falling in area of a geometric shadow for a divergent flow is considerably greater then that in the case of a collinear flow. One may say that the distance a between the collector and the edge of diaphragm efficiently decreases.

We shall note one important circumstance. The scattered flow is proportional to the squared vapour density in the ionisation zone. Thus, using the results of simulations and measurement of a scattered flow density we can determine the mean vapour density in the ionisation zone.

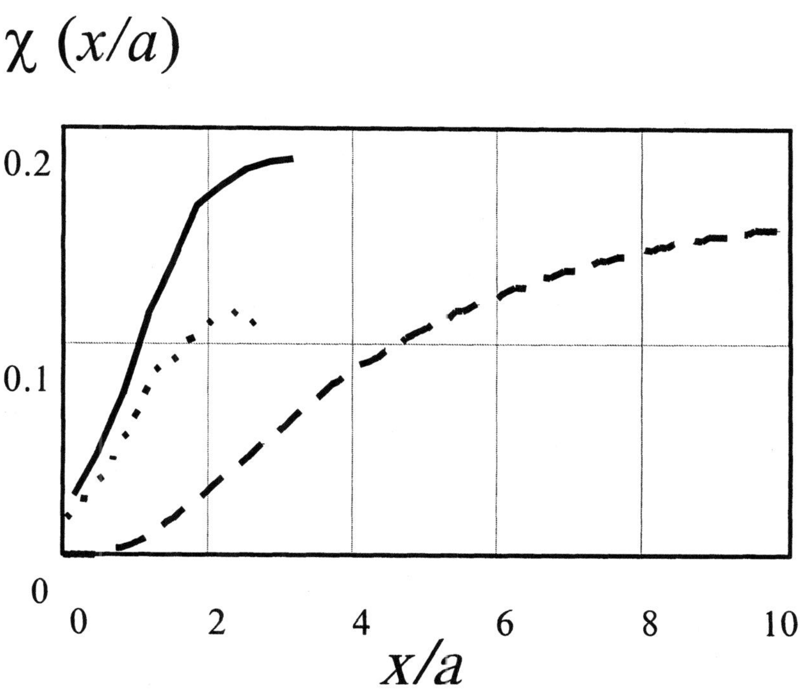

Figure 12: The scattered flow due to of single collisions. The function $\chi(x/a)$ gives the probability density that an atom falls within unit area of the plane parallel to the vapour flow via the distance x from the diaphragm. Various curves correspond to a different distances from symmetry axis: (*1*) 2 cm; (*2*) 4 cm; (*3*) theoretical dependence for collinear atom flow

4.3 Influence of atom collisions on the properties of flow

Investigated problems.

The statistical 'particle-in-a-box' method was used to deal with the problems of the passage of atoms flowing through a slit [15], of the flow of atoms into vacuum from a slit, and of planar flow in vacuum [16]. The first case deals with the flow of atoms formed in an infinitely long slit with diffusely reflecting walls. The x-axis is directed along the slit and the y axis is across the slit; the slit is assumed to be infinite along the z-axis. The condition of specula reflection is used on the symmetry axis ($y = 0$).

When the flow of atoms from a slit into vacuum is considered, it is assumed that atoms do not emerge from the whole area at the bottom of the slit, but from a strip of width d less than the width of the modelled region. The distribution of atoms evaporated from such a strip is given by expression (5). It is assumed that an atom, which reaches the edge of the modelled region, disappears there. The specula reflection coefficient is again used on the symmetry axis. The same condition is specified also when modelling planar evaporation on slit walls.

Angular distribution of particles escaping from a slit.

Calculations show that the influence of collisions on the angular distribution of the escaping particles is important even when the length (and not the width) of the slit becomes equal to the mean free path (Fig. 13). A peak of the distribution in the low-angle range decreases considerably even if $l\sigma N_0 \approx 0.3$ (the transverse size in units of the mean free path is very small: $\Delta l\sigma N_0 \approx 0.05$). However, when the mean free path becomes comparable with the slit length, we have $l\sigma N_0 > 2$ and the angular distributions are practically identical in planes perpendicular and parallel to the slit (and passing through its centre).

In this connection we should point out the following circumstance. The angular distribution of particles escaping from a narrow channel is frequently considered in the following simplified manner. The channel length is effectively reduced to the mean free path and the subsequent motion of particles is assumed to be free-molecular. The results of our calculations show that the approximation is invalid at least in the two-geometry (slit) case. It is therefore desirable to assume that the distribution corresponds to the distribution of particles emitted by an equilibrium source.

Planar evaporation in a vacuum.

In this problem the greatest interest lies in the flow which is formed or in the fraction of the particles which are returned. The flow density j_x can be measured conveniently in terms of the density of flow of the injected particles (with $\mathbf{V} = 0$):

$$j_x = 2\sqrt{\pi}\,\frac{v_x}{v_T} \cdot \frac{N(x, y)}{N_0}\,.$$

Here, N_0 is the vapour density at the boundary which emits atoms; $v_T = (2T/M_{\mathrm{Yb}})^{1/2}$.

Calculations show that when the distance to the boundary where particles disappear is considerably less than the mean free path, the flow density is $j_x = 0.815 \pm 0.005$. In other words, the fraction of the particles returned to the surface is 18%-19%.

Calculations of the density of atoms in the ionisation zone.

In dealing with the problem of laser separation of isotopes the main practical interest lies in calculation of the distributions of the vapour density and of the temperature in the ionisation zone. In the experiments described in Refs. [1-3], the vapour source was a system of guiding channels (for details see Ref. [15]). In this case the average velocity of the injected atoms was not zero ($\mathbf{V} \neq 0$). Experiments involving measurements of the total rate of flow of the evaporated ytterbium and of the amount of ytterbium passing through an aperture showed that the best agreement with calculations for such a system of guiding channels is obtained on the assumption that $V = 1.4 \cdot v_T$. The results of calculations of the distributions of the various characteristics over the transverse cross section of the flow are plotted for this case in Fig. 14. The data on vapour densities were inspected by independent measurements of a scattered flux.

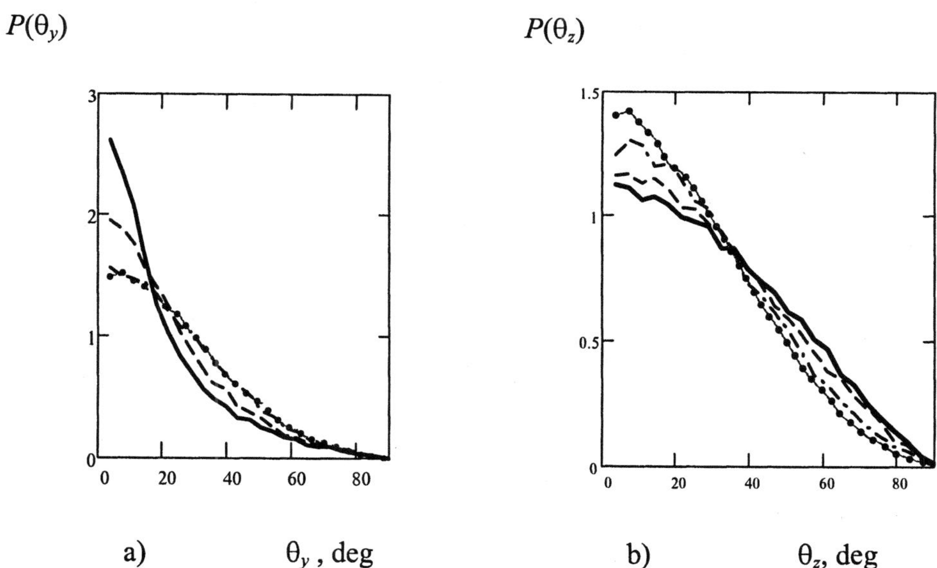

Figure 13: Angle distributions of atoms transmitted throw the split: à) on an angle θ_y in a plane perpendicular split; b) on a angle θ_z in a plane parallel split (through its middle):
1) the solid curves correspond to ignoring of collisions;
2) the dashed curves correspond to density of vapours on an slit entrance $N_0 = 3 \cdot 10^{13}$ cm$^-$^{3}, $l\sigma N_0 \approx 0.32$, $\Delta l\sigma N_0 \approx 0.03$ (l - height of a split, Δl - width of a split, σ - cross section of atom - atom collisions);
3) the dot-dash curves correspond to $N_0 = 2 \cdot 10^{14}$ cm^{-3}, $l\sigma N_0 \approx 2.2$, $\Delta l\sigma N_0 \approx 0.2$;
4) the dotted curves correspond to $N_0 = 6 \cdot 10^{14}$ cm^{-3}, $l\sigma N_0 \approx 6.5$, $\Delta l\sigma N_0 \approx 0.55$;
5) the thin curves correspond to $N_0 = 1.2 \cdot 10^{15}$ cm^{-3}, $l\sigma N_0 \approx 13$, $\Delta l\sigma N_0 \approx 1.1$
(the curves 3,4,5 on Fig 13a have practically merged)

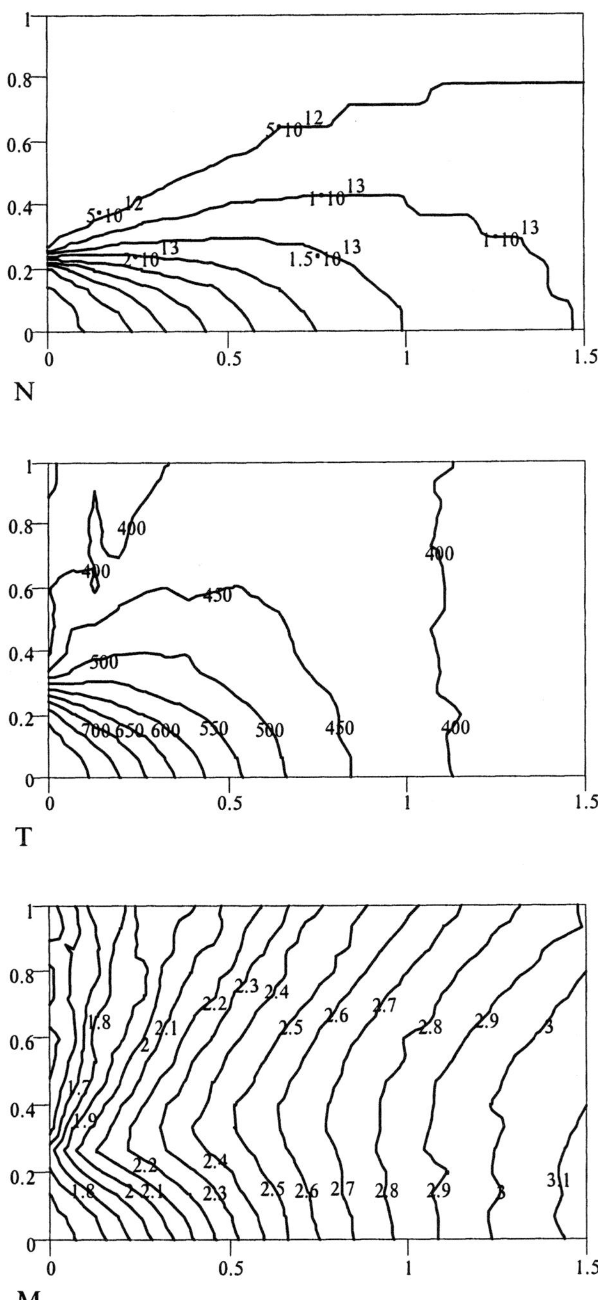

Figure 14: Contour lines of various gas parameters, along a vapour flux flowing out into vacuum: $N(x,y)$ is the atoms density; $T(x,y)$ is the gas temperature in ^{0}K ; $M(x,y)$ is the local Mach.

The calculation was done for the 100 gm/hr Yb vapour consumption and for $d = 0.5$ cm. The initial densities of atoms, temperature and average velocity of flowing out vapours were assumed as follows: $N_0 = 6 \ 10^{14}$ cm^{-3}; $T = 850 \ {}^0$K; $V = 1.4 \ v_0$ ($V = 2.86 \ 10^4$ cm/s).

5 Extraction of ions

Free expansion.

There are various methods for ion extraction from plasma formed as a result of selective multistage ionisation. It is usual to employ electric and magnetic fields, as well as their combinations. For a variety of reasons we shall consider the use of an electric field created by a system of electrodes.

Under the investigated conditions the role of the space charge is very important. For example, if plasma placed between the electrodes of a capacitor, which are subjected to a high-voltage pulse, electrons are extracted from the plasma and a Coulomb explosion of an uncompensated ion charge occurs [18].

The main feature of the plasma formed by the AVLIS method is that the mean free path of particles in this plasma is much greater and the Debye radius is much less than the transverse dimensions of the ionisation region. A numerical solution of the equations describing 'two-fluid' hydrodynamics was used in Ref. [19] to show that, under conditions of steady-state expansion, the average-mass velocity is equal to about half the most probable velocity of the particles that have the temperature of electrons and the mass of ions; such plasma expansion is established in a time approximately 1.5 times longer than the time needed for traversing the initial transverse dimension at the steady-state expansion velocity. This means that it is reasonable to expect efficient ion extraction by free expansion if potential contamination of the ion collector can be ignored.

Steady-state two-dimensional model.

The problem of ion extraction is complicated by the fact that even in a low-density gas the collector may be strongly contaminated by single collisions of atoms in a gas jet. It is therefore of interest to consider a collector in the form of grids creating fields, which accelerate ions and decelerate electrons. Such grids can also protect the collector from neutral particles. In particular, it is possible to collect ions on the surfaces of the grid components located in the region of the geometric shadow of the scattered flux of atoms.

The transparency of the grid cells was investigated as a two-dimensional steady-state problem [23]. An analysis was made of the motion of ions in a self-consistent electric field formed both by electrodes and by space charges appearing in plasma because of a departure from quasineutrality. Electrons were assumed to have the Boltzmann distribution with the temperature T_e. Since the motion of ions between the electrodes was fairly complex, the solution of the hydrodynamic equations in combination with the Poisson equation was a difficult computational task.

Use was made of the particle method together with iterations of the Poisson equation in terms of the ion density. During the initial stage selecting the electrode potentials specified the electric field. This was followed by an analysis of the motion of a large number of probe ions in this field and the spatial distribution of the ion density was determined. This distribution, in combination with the Boltzmann distribution of electrons, determined the space charge density at various points in space. The Poisson equation was solved for the space charge density obtained in this way and a new spatial distribution of the field was found. The motion of probe ions in this field was considered again, and so on. The calculation method described above was performed with the aid of a 'Potentsial' suite of programs. Given positions and electrode potentials and parameters of the plasma flowing into the region under investigation were used to calculate the spatial distributions of the particle density and of the potential. Moreover, the number of particles reaching some particular surface in the investigated region or on the electrodes was calculated. The calculations were carried out by the method for solving the Poisson equation described in Ref. [95].

Examples of the results of such calculations are plotted in Figs 15 and 16. They illustrate the situation when an electric field is generated by a potential difference between two grids formed by strips 6 mm wide, infinite parallel to the z direction and separated by a distance of 6 mm from one another along the y axis. Only a part of the region was modelled and the full picture was obtained by symmetric continuation of the results of calculations along the y axis. The field was assumed to be created by two rectangular electrodes, defined by co-ordinates of the left (x_l, y_t) and right bottom (x_r, y_b) corners, and by the potential φ. Ions were supposed to arrive through the left boundary of the investigated region at an average velocity $v_0 = 3 \cdot 10^4$ cm s^{-1}, directed parallel to the x axis. Modelling was carried out for 500 particles and five density iterations were carried out.

At sufficiently high voltages between the grids the electric field penetrates the region in front of the first grid kept at zero potential and drives electrons out of this region (Figs 15a and 16a). Beginning from a certain voltage the pressure exerted by the field exceeds the electron pressure even opposite the centre of the first grid, so that electrons cannot approach the electrodes at all. This is precisely the situation assumed in one-dimensional models discussed below.

However, the electric field does not alter too much the number of ions reaching the first grid. This is clear from the distributions of the ion density (Figs 15b and 16b). The calculated results show that at a low voltage between the grids (10 V) the fraction of ions reaching the first electrode is close to that given by the geometric probability (50.8%), whereas at high voltages (1 kV) it falls only by a factor of 1.8 (to 28.8%). Therefore, when the aim is to increase the grid transparency, one should first reduce the relative size of the electrode area compared with the size of the aperture, rather than try to increase the voltage.

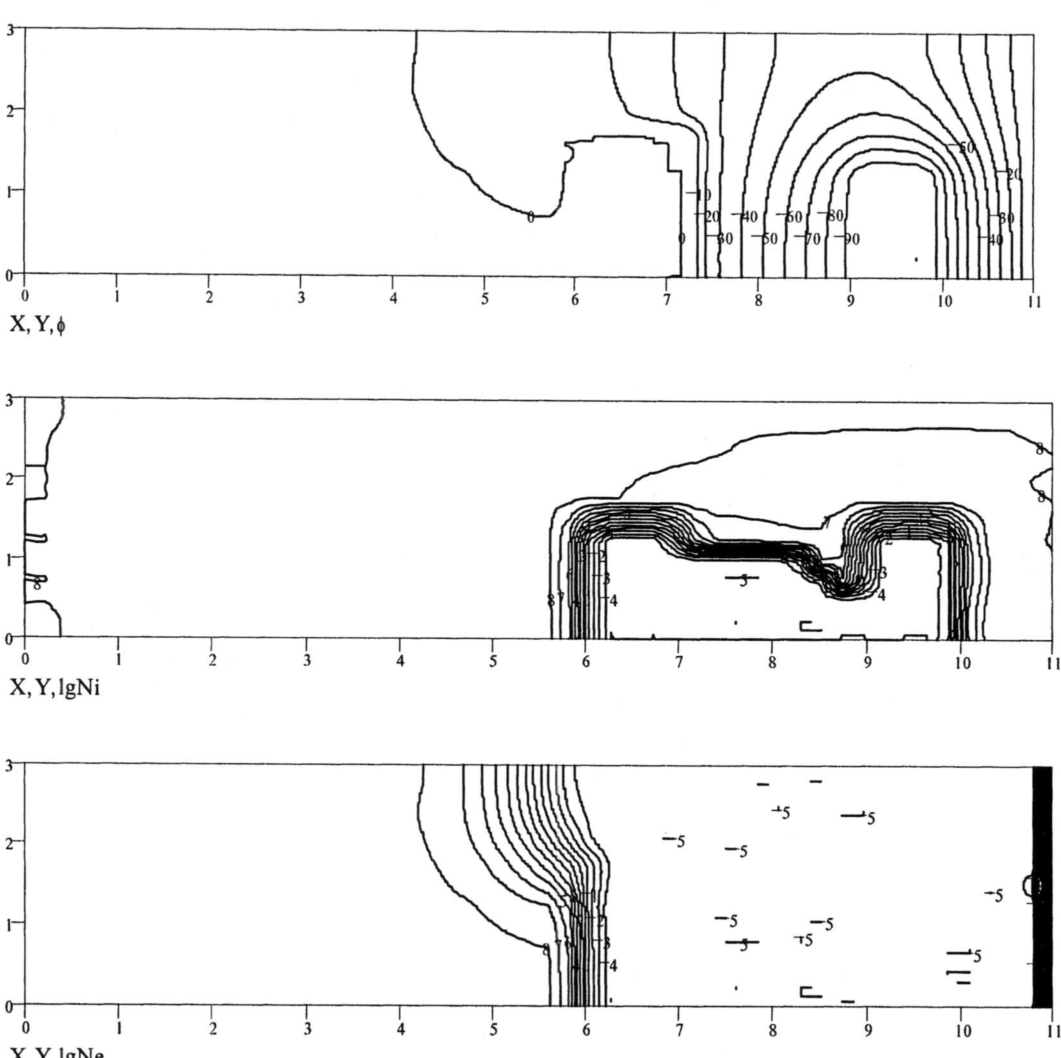

Figure 15: Contour lines for one value of voltage between grids: a) field potential; b) the logarithm of ions density; c) the logarithm electrons density. The width of calculations area is 11 mm (along an axis X); a height of area is 3 mm (along an axis Y).

The field forms by two rectangular electrodes (assigned co-ordinates of left upper x_l, y_t and right lower x_r, y_b angles, and also potential φ). The first electrode: $x_l = 6$ mm, $y_t = 1.5$ mm; $x_r = 7$ mm, $y_b = 0$; $\varphi_1 = 0$. The second electrode: $x_l = 9$ mm, $y_t = 1.5$ mm $x_r = 10$ mm, $y_b = 0$, $\phi_2 = -100$ V (voltage between grids $U_0 = -100$ V). Ions density on the left boundary is $N_{e0} = 10^8$ cm^{-3}; temperature of electrons is $T_e = 0.17$ eV; temperature of ions is $T_i = 0.03$ eV. The ions pass through the left boundary of the calculations area with the average velocity $v_0 = 3 \cdot 10^4$ cm/s, directed in parallel axes X. During simulations the trajectories 500 particles were calculated; five iterations on a denseness were carried out. The grid 49×12 was used.

N_e, cm^{-3}

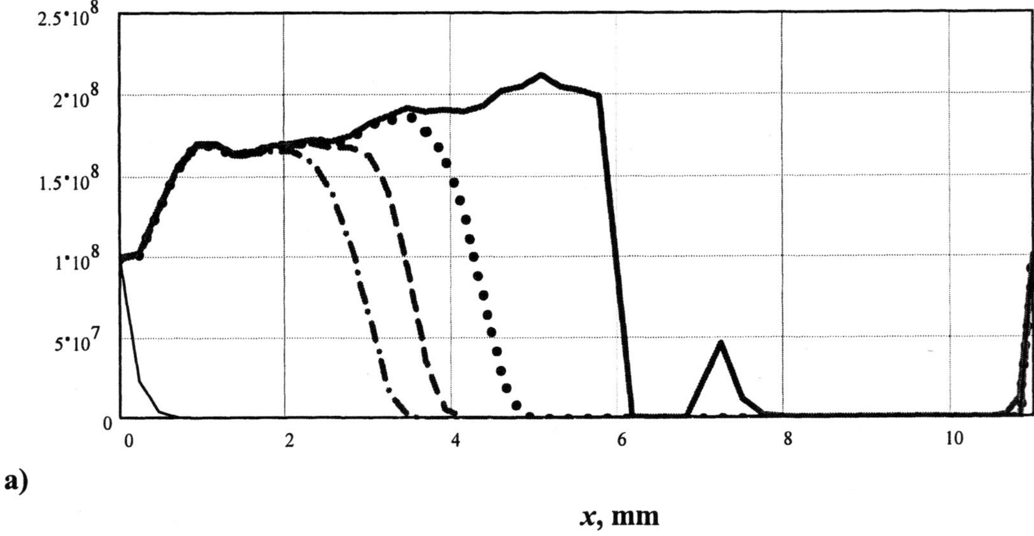

a)

x, mm

N_i, cm^{-3}

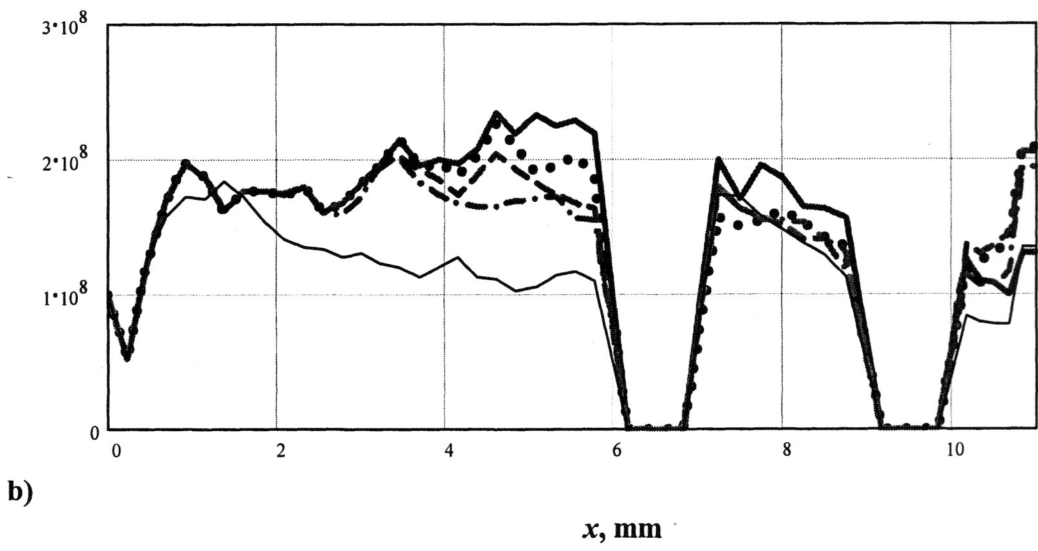

b)

x, mm

Figure 16: Distributions on an axis X for $y = 0$ for different values of voltage between grids: a) electrons densities; b) ions densities. Solid lines correspond to $U_0 = -10$ V; dotted lines correspond to $U_0 = -100$ V; dashed lines correspond to $U_0 = -200$ V; a dot-dash line correspond to $U_0 = -300$ V; a thin solid correspond to $U_0 = -1000$ V. Other parameters are the same, as for Fig. 14.

Time-dependent one-dimensional models.

Expansion [19] and passage of plasma through grids [20 - 22] were considered on the basis of a time-dependent one-dimensional two-fluid model. The grids were regarded as transparent and the potentials applied to them were described by introducing additional charges into the appropriate points in space (Fig. 17). Problems of this kind are traditional in physical electronics [119-121] and it would seem they have been thoroughly investigated. However, an analysis of the behaviour of a laser plasma detached from electrodes predicts effects which had not been given sufficient attention earlier.

For example, it is usual to assume that if a collisionless plasma is in contact with parallel grids to which a potential difference is applied, the result is extraction of charged particles from the plasma and the density of the corresponding ion or electron current is then described by the familiar Child's equation leading usually to the 'three-halves' law [119-121]. However, this actually occurs only if the charge is not retained in the investigated space, i.e. if there are charge sources and sinks (usually at the boundaries). However, if (for example) a plasma is surrounded by a double layer of grids from all sides so that the charge can escape only through such a double layer, the application of a potential difference to the grids locks in not only particles of one sign, but also nearly all the plasma. Moreover, as shown in Ref. [22], even if the electron charge disappears at some electrode outside the plasma, but the electron-extracting potential is not applied to this electrode, the passage of plasma through transparent grids is also hindered.

This effect is illustrated in Fig. 18. Dimensionless quantities are used in this figure. The units of the dimensionless quantities were expressed in terms of the maximal initial density and pressure n_0, p_{e0}; the initial temperature is then $T_0 = p_{e0} / n_0$. The lengths (distances) were measured in terms of the Debye radius $l_0 = r_{\mathrm{D}} = \sqrt{T_{e0} / 4\pi e^2 n_0}$; the time in terms of inverse Langmuir frequency $t_0 = \omega_{\mathrm{L}}^{-1} = \sqrt{m_e / 4\pi e^2 n_0}$; densities in units of maximal initial density of electrons and ions n_0 (plasma at the initial time moment is quasineutral); velocities in units of mean thermal velocity of electrons $v_0 = \sqrt{T_{e0} / m_e}$; potential and field intensity were expressed in terms of the initial electron temperature: $\varphi_0 = T_{e0} / e$, $E_0 = T_{e0} / el_0$. The simulations were done for a not very small mass ratio $\mu = m_e/M_i = 0.01$, were M_i is the ion mass.

It was shown, that the locking of ion current and reflection of plasma from grids take place if plasma placed between zero potentials (Fig. 18a). Thus condition of a strong breaking plasma field between grids is not enough for transportation of ions through a grid. To achieve a sharp increase of an efficiency of the ions extraction by weak fields it was proposed to use a removed electrode under a low potential pulling out the electrons (see Fig. 18b).

Let us assume that, for example, $N_e \sim 10^8$ cm^{-3} and $T_e \sim 0.1$ eV. We then have $r_{\mathrm{D}} \sim 2$ 10^{-2} cm and $T_e/r_{\mathrm{D}} \sim 5$ V/cm. It is natural to consider typical dimensions of plasma and assume that the distance to the nearest grids is of the order of 1 cm. It is then found that

$x_0 \sim x_S \sim 50$ (see Fig. 17). The distance between the grids is naturally assumed to be of the order of several millimetres, so that we have $x_0 - x_A \sim 10$. The physical time τ for the ion component can be found by multiplying the dimensionless time t by the factor $\beta = (1/\omega_L)/(M_{Yb}\mu/m_e)^{1/2}$ (where $\mu = 0.01$ is the ratio of the masses of the negative and positive particles used in the calculations). For the parameters assumed above, we have $\beta \sim 100$ ns. A typical ion-extraction time is ~ 50 μs if the parameters are those assumed in Fig. 18. However, the efficiencies of extraction with and without an extracting electrode are very different: they are 73% and 8%, respectively. When the velocity of the Yb vapour through the selective ionisation zone is $\sim 3 \cdot 10$ cm s^{-1}, the vapour stream is displaced by $\sim$ 1 cm along the grid plane during the extraction time.

These physical effects were taken into account in the construction of an extractor for the isotope separation facilities described in Refs. [1-4]. In particular, a specific extractor configuration was selected after repeated modelling of a number of different variants of the geometry of the accelerating electrodes and of the collector. This made it possible to ensure efficient (30%-70%) ion extraction with very slight contamination of the collector by the flow of neutral particles.

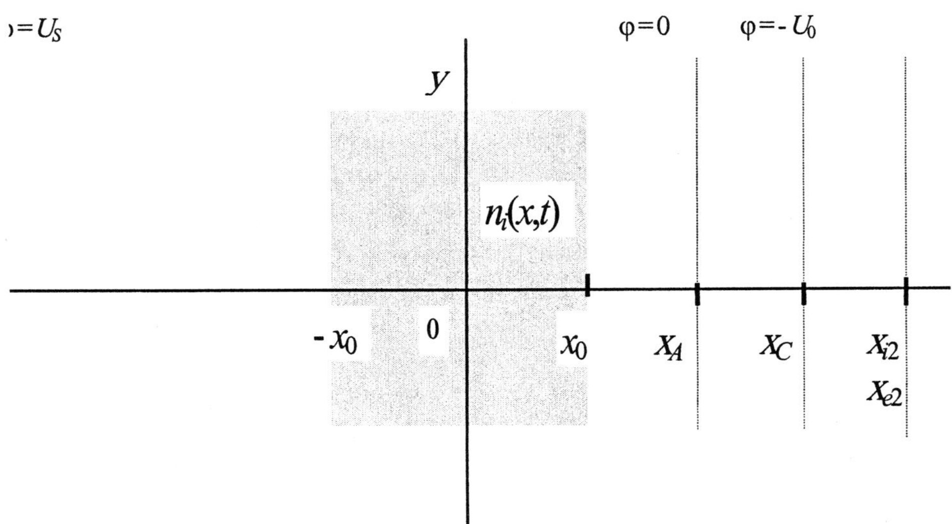

Figure 17: Geometry of an expanding plasma layer, the grids, the collector and the vacuum chamber walls: $x_0 = 20$ is the initial half-width of a plasma layer; $x_S = x_{i1} = x_{e1}$ are the co-ordinates of the first grid S under a voltage U_S and the first point, in which charges of ions and electrons are lost (the wall of the vacuum chamber); $x_A = 25$, $x_C = 40$ - co-ordinates of the grids A and B, between which the voltage $-U_0$, extending ions, is applied; $x_{i2} = x_{e2} = 45$ is the second point, in which charges of ions and electrons are lost (the collector).

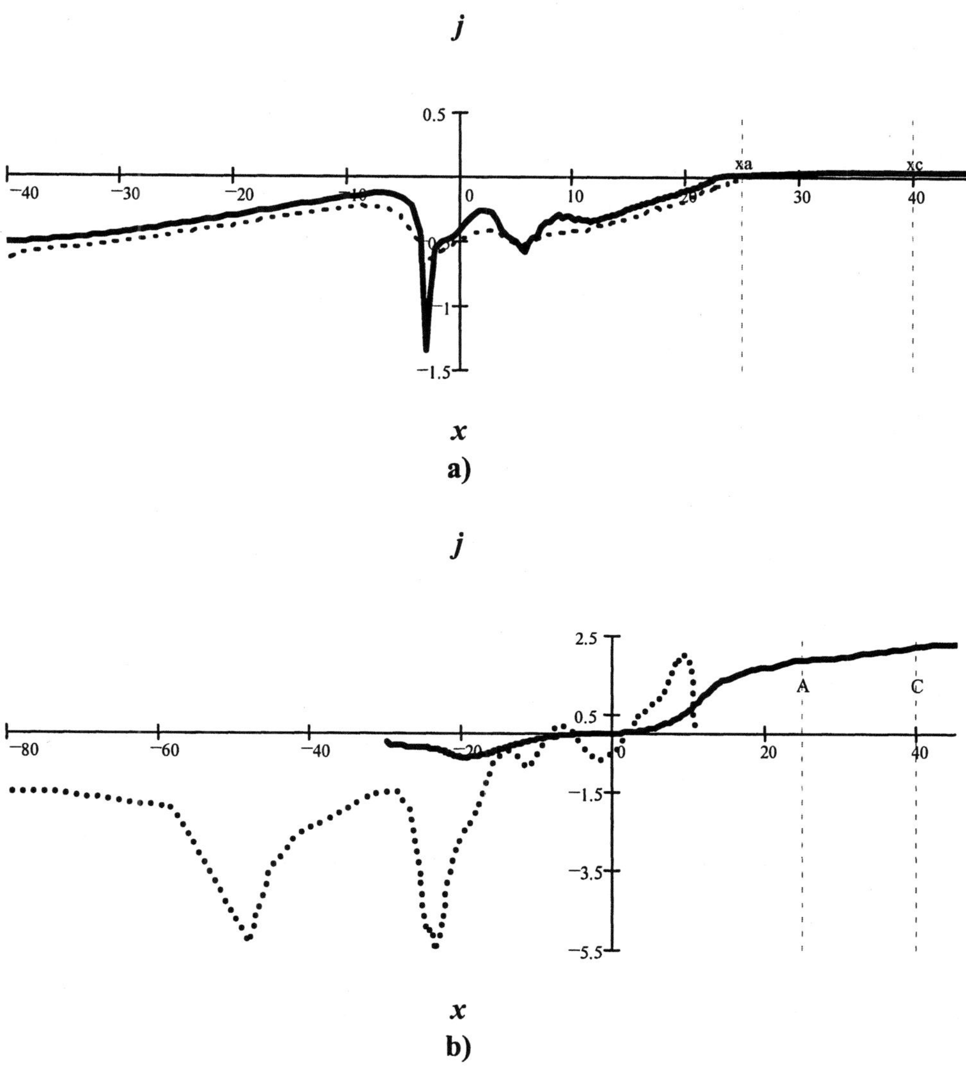

Figure 18: Space distributions of the density of ionic flow j_i (solid lines) and electronic flow j_e (dotted lines):
a) for the case of plasma placed between two grounded grids ($x_S = x_{i1}, = x_{e1} = -40$, $x_0 = 20$, $x_A = 25$, $x_C = 40$, $x_{i2} = x_{e2} = 45$, $U_0 = 100$, $U_S = 0$, $\mu = m_e/m_i = 0.01$, $d_0 = 2$) at time $t = 330$. The plasma was reflected from double layer of transparent grids;
b) for the case of plasma under the potential pulling out electrons ($x_S = x_{i1}, = x_{e1} = -80$, $U_S = 100$, at the time moment $t = 63.5$, other parameters are the same as for the case a). The ions have penetrated through grids when electrons were extracted

6 Production of a highly enriched isotope

6.1 Production facility

Principal parameters.

Without going into details we shall now list the basic parameters which we were guided as the result of the first stage of computations and experiments:

- a vapour density $(1\div5)\ 10^{12}$ cm^{-3} in an ionisation zone $1\div2$ m long with cross section area $5\div10$ cm^2;
- average dye laser powers at a pulse repetition frequency 10 kHz of 1 W at the wavelength $\lambda_1 = 555.6$ nm, 1 W at $\lambda_2 = 581.1$ nm, and $3\div5$ W at $\lambda_3 = 582.8$ nm.

These parameters determine largely the nature of created apparatus [1,2]. Fig. 19 shows schematically the basic layout of the installation.

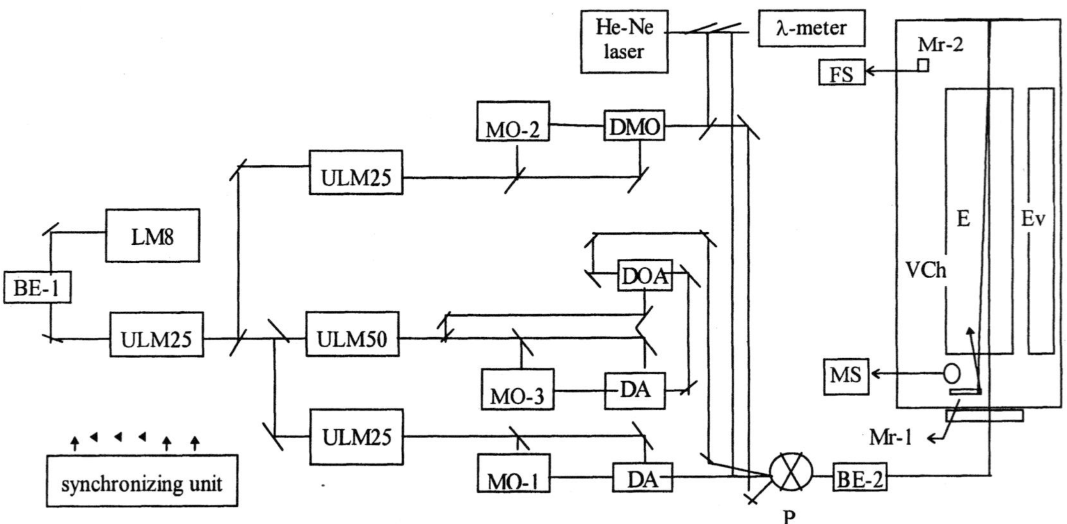

Figure 19: Schematic diagram of the system (installation) used in laser separation of ytterbium isotopes in weighable amounts: BE - beam expander; P - periscope; Mr - mirrors; E - ion extractor; FS - flux sensor; VCh - vacuum chamber; LM and ULM - copper-vapor lasers; MO - copper-vapor master oscillator; DMO - dye master oscillator; DOA - dye output oscillator; MS - mass spectrometer

Copper and dye laser system.

The dye lasers were pumped by a system of copper-vapour lasers LM8, ULM25, ULM50 generating pulses of 15 ns half-width at repetition frequency 10 kHz. The first of these lasers was used as the master oscillator and others were employed as amplifiers. The multichannel electronic unit provided synchronisation of the laser pulses to within $1 \div 2$ ns. The average output power at the 510 nm was 18 and 30 W when the ULM25 and ULM50 amplifiers were used.

The dye laser system had also three channels with the master oscillator-amplifier configuration (MA, DA). The output power of the third channel ensuring photoionisation was increased by the use of a second amplifier (DOA).

The master oscillator determined the spectral composition of the dye laser radiation where the main spectrally selective components were a diffraction grating and a Fabry-Perot etalon. The laser wavelength was monitored with the aid of a wavelength (λ–) meter calibrated with a stabilised He-Ne laser. Moreover, the radiation quality was monitored on the basis of the ion isotopic composition of the laser plasma determined with the aid of a mass spectrometer (MS).

Two type rhodamine dyes were used: R6G and R110. The laser channel efficiency was $20 \div 25$ % without the etalon in the master oscillator and $15 \div 18$ when the etalon was present. The half-width of the laser lines at the first and second step of excitation were measured with Fizeau and scanning Cann interferometers: it did not exceed 0.5 GHz (see section 3.2). The short-term instability of the lasers wavelengths was 200 MHz, and long-time instability, due to change of temperature was 400 MHz/h. During long operation (up to 16 hours) spectral and power characteristics were reproduced rather well. The frequency drift, caused by temperature variation was corrected every 0.5 hour.

At the end of 1997 a larger facility was constructed with a 4.5-m long active length of the vacuum chamber (Figs 20-22) and the powers of both the copper and dye lasers were increased. In all cases the three-stage amplification was employed.

System for combining the laser beams and forming the photo-ionisation zone.

One of the main factors governing the photoionisation efficiency and the efficiency of the whole isotope production process proved to be selection of the optimal system for combining the laser beams and forming the photoionisation zone inside an atomic beam generated by an evaporator in a vacuum chamber. The optimisation criterion was full utilisation of each photon. The difficulties encountered in constructing such a system arose from a number of causes, including the spatial and angular distributions of the three laser beams, and the dependence of the volume of the ionisation zone on the vapour density, on the laser radiation powers, and on other factors.

The system under consideration could be regarded as divided arbitrarily into two parts. The first part ensured equalisation, by a set of collimators, of the transverse cross sections of all three-laser beams. It then formed (by a set of segmented mirrors, a dichroic mirror, and a mirror telescope) a combined beam in which the component beams were made to coincide over a distance in excess of 1 m. The second part of the system had a multipass configuration consisting of a mirror for injecting the combined beam and of optical cavity mirrors set in the vacuum chamber windows.

The combined beam was injected into the cavity so as to ensure the largest possible number of reflections between the mirrors. The volume of the ionisation zone then depended on the number of reflections and on the magnification of the mirror telescope, as well as on the positions of the collimators that determined the single-pass cross section of the combined beam inside the chamber.

Careful alignment made it possible to achieve reproducibility up to 40-50 reflections, each of which increased proportionally the contribution to the photocurrent in the extractors (see Section 3). The use of such a system made it possible to vary quite flexibly the dimensions of the ionisation zone and to reach much higher photocurrents than in the single-pass configuration.

Figure 20: Photograph of a dye laser system

Figure 21: External appearance of the vacuum chamber

Figure 22: Photograph of the concurrent laser beams put into the vacuum chamber

Intrachamber components.

The chamber contained an evaporator, a system of extractors, sensors for monitoring the vapour density, and a mass spectrometer (Fig. 23).

The evaporator operated on the basis of the Joule heating process, ensuring generation of an atomic vapour of density up to 10^{13} cm^{-3} in the photoionisation zone. This could be maintained for many hours. The power consumption was up to 3 kW. In most of our experiments the vapour source was a slit 1 m long and 2 mm wide. In some experiments extended tubes or similar guiding components formed an atomic beam. Prolonged operation of the evaporator demonstrated its high reliability, the absence of blocking of the slit, and a good reproducibility of the atomic beam parameters.

One of the important intrachamber components was the extractor system because its operational efficiency determined the efficiency of the facility as a whole. The selected extractor design was of the venetian-blind type. This had a number of advantages such as the high efficiency of ion collection, a low background contamination level, and the ability to operate in a wide range of voltages (from 10 V to 4000 V).

The degree of enrichment of the mixture with the target isotope was monitored by a permanent-magnet mass spectrometer, operating on-line with a personal computer.

J (rel. units)

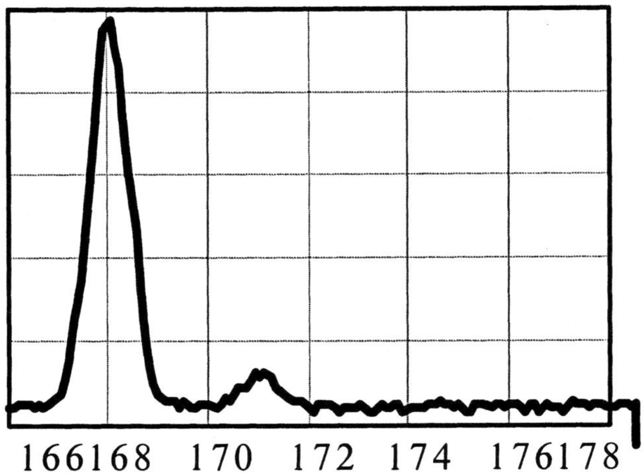

Figure 23: Mass-spectrometer current as a function of atomic mass. Example of the isotope composition in the laser produced plasma (^{168}Yb concentration is 93 %)

Figure 24: A working moment of meeting at the 'Lad' Scientific-Production Enterprise (1995).
From left to right V I Derzhiev, P I Gromov, S I Yakovlenko, V M Mushta

6.2 Experimental results

Many series of experiments were carried out on this facility and they involved investigations of the formation of the vapour beam and of the dependence of the ion composition of the photocurrent on the wavelengths, on the spectral compositions, and on the powers of the laser beams used in the various excitation stages, and also on the configuration of the zone exposed to the laser radiation.

The main experiments were carried out under the following conditions:

—the width of the laser emission line used in the first and second stages was ~ 500 MHz;
—the output powers were 1,1, and 4 W for the first, second, and third lasers, respectively;
—the vapour density was varied in the limits $10^{11} \div 10^{13}$ cm"^{3};
—the cross section of the ionisation zone was varied from 1 to 4 cm^2;
—the working voltages on the extractors were varied within the range 100-3800 V.

Some of the experimental results can be summarised as follows.

1. The maximum attained degree of enrichment was 32% and the mass spectrometer recorded enrichments of 80%-95% (Fig. 23). The degree of enrichment was reduced by the background contamination of the collector plates in the extractor. Charge transfer did not affect the degree of enrichment throughout the investigated range of the atomic vapour densities.

2. It was found that the frequency stability of ±350 MHz in the first stage and ±200 MHz in the second stage was essential to ensure a high degree of enrichment.

3. The extraction efficiency depended on the applied voltage and it increased nonlinearly (reaching saturation) when the voltage was increased. One extractor did not influence the other one during operation. The photocurrents from the separate extractors were summed and reached ~ 100 μA in these experiments.

4. In the course of these experiments we tested methods for minimising the background contamination with the aid of traps designed in a number of ways. An increase in the vapour density increased the photocurrent nonlinearly, but it reduced strongly the content of the target isotope on the collector.

5. Our experiments demonstrated the importance of the system for combining the beams and for forming the ionisation zone. Optimisation of the parameters of this system made it possible to reach values of the photocurrent close to those predicted by calculations.

Our experimental results (the ion current and the enrichment with Yb) were in good agreement with the results of a numerical simulation carried out in a wide range of the input radiation parameters (power, spectral composition geometry of the laser beam injection). All the experimental results could be interpreted correctly by theoretical calculations.

Some difficulties were encountered in modelling the processes of scattering by components of the system, giving rise to additional background contamination. It was extremely difficult to include the structural factors in any analysis, so those simple models were employed [15]. Recommendations were developed for constructing efficient vapour traps, which reduced greatly the contamination background.

In 1996-97 the ion current to the collector reached 0.5-1.5 mA per 1 m of the active medium, which corresponded to about 3-10mg/h of the enriched ^{168}Yb from the 1 m length. In the course of a month we were able to operate the facility for over 200-h and produced approximately 0.5-2 g of the isotope. The results of independent mass-spectrometric analyses, carried out at the State Scientific-Research and Design Institute of the Rare-Metal Industry in Moscow, indicated that in the material deposited on the collector the ^{168}Yb content was ~ 62% and in the washing liquid it was 45%.

7 Conclusions

The work on laser separation of isotopes, carried out at the Institute of General Physics of the Russian Academy of Sciences and at the 'Lad' Scientific-Production Enterprise, involved theoretical studies, computer simulation, and experimental investigations of the processes of separation of the ytterbium isotopes by the AVLIS method.

These investigations made it possible to solve the key problems of laser separation of weighable amounts of an isotope when its abundance in a natural mixture was low. These problems included:

(1) ensuring a high selectivity of the ionisation in a relatively large volume of an optically dense vapour (it was proposed to increase the ionisation efficiency by a suitable distribution of the laser beams in a cavity and this was implemented successfully);

(2) formation, over a distance 1 m, of a homogeneous vapour beam which would not contaminate the collector;

(3) efficient extraction of ions from the plasma, avoiding parasitic ionisation of the vapour and preventing contamination of the collector with the vapour characterised by the natural abundance of the isotopes.

These investigations were the basis for constructing of facility capable of producing highly enriched ^{168}Yb on an industrial scale. The wide use of the numerical simulations has essentially reduced the costs of preliminary experiments and optimisation process.

These results represent a qualitative leap forward in the task of laser isotope separation by the AVLIS method. The cost of highly enriched (in excess of 25%) ^{168}Yb is high because its abundance in the natural mixture is very low. Consequently, even production of 1 g of ^{168}Yb per month is profitable. Isotope separation by AVLIS method not on a demonstration but on a commercial level was carried out for the first time. Or else, for the first time profitable commodity was produced by AVLIS method. One can expect development of industrial production also of other isotopes for medical purposes.

Acknowledgements

V I Derzhiev and V M Mushta were responsible for the financial aspects and for the engineering and organisational control of the work. They showed great courage in taking up this task under the current unpredictable conditions in Russia. I am grateful to my colleagues R I Golyatina, P I Gromov, A G Egorov, A A Il'in, S A Kostritsa, V A Kuznetsov, L A Mikhal'tsov, A A Polovtsev, A Yu Sapozhkov, A N Trachev, S A Chaushanskii for responsible participation in this series of investigations.

V I Mishin has attracted my attention to AVLIS topic and introduced me into the heart of the matter. V I Mishin suddenly have died in December 1998. The colleagues will save light memory of this uncommon person.

References

[1]   Derzhiev V I, Kuznetsov V A, Mikhal'tsov L A, Mushta V M, Sapozhkov A Yu, Tkachev A N, Chaushanskii S A, Yakovlenko S I *Kvantovaya Elektron. (Moscow)* **23** 771 (1996) [*Quantum Electron.* **26** 751 (1996)]

[2]   Derzhiev V I, Kuznetsov V A, Mikhal'tsov L A, Mushta V M, Sapozhkov A Yu, Tkachev A N, Chaushanskii S A, Yakovlenko S I *Proceedings of the International Conference on Lasers 96, Portland, OR, 1996* (McLean, VA: STS Press, 1997) p. 441

[3]   Derzhiev V I, Egorov A G, Il'in A A, Kostritsa S A, Kuznetsov V A, Mikhal'tsov L A, Mushta V M, Polovtsev A A, Sapozhkov A Yu, Tkachev A N, Chaushanskii S A, Yakovlenko S I *Dokl. II Vseros. Konf. Tiziko-Khimicheskie Protsessy pri Selektsii Atomov i Molekul'* (Zvenigorod 1997) (Paper presented at the Second All-Russia Conference on Physicochemical Processes in Selection of Atoms and Molecules, Zvenigorod, 1997) (Moscow: TsNIIatominform, 1997) p. 57

[4]   Derzhiev V I, Kostritsa S A, Kuznetsov V A, Mikhal'tsov L A, Mushta V M, Sapozhkov A Yu, Tkachev A N, Chaushanskii S A, Yakovlenko S I *Proc. SPIE Int. Soc. Opt. Eng.* **3403** 242 (1998)

[5]   Derzhiev V I, Kostritsa S A, Kuznetsov V A, Mikhal'tsov L A, Mushta V M, Sapozhkov A Yu, Tkachev A N, Chaushanskii S A, Yakovlenko S I *Kvantovaya Elektron. (Moscow)* **25** 287 (1998) [*Quantum Electron.* **28** 278 (1998)]

[6]   Letokhov V S, Author's Certificate No. 784 679 appl. 30 March 1970, in *Byull. Izobret.* (8) 308 (1982)

[7]   Krynetskil B B, Mishin V A, Prokhorov A M *Zh. Prikl. Spektrosk.* **54** 558 (1991)

[8]   Tkachev A N, Yakovlenko S I *Kvantovaya Elektron. (Moscow)* **23** 860 (1996) [*Quantum Electron.* **26** 839 (1996)]

[9]   Tkachev A N, Yakovlenko S I *Kvantovaya Elektron. (Moscow)* **24** 759 (1997) [*Quantum Electron.* **27** 740 (1997)]

[10]   Derzhiev V I, Kostritsa S A, Kuznetsov V A, Mikhal'tsov L A, Mushta V M, Sapozhkov A Yu, Tkachev A N, Chaushanskii S A, Yakovlenko S I *Dokl. II Vseros. Konf. Tiziko-Khimicheskie Protsessy pri Selektsii Atomov i Molekul' (Zvenigorod 1997)* (Paper presented at the Second All-Russia Conference on Physicochemical Processes in Selection of Atoms and Molecules, Zvenigorod, 1997) (Moscow: TsNIIatominform, 1997) p. 83

[11]   Golyatina R I, Tkachev A N, Yakovlenko S I *Kvantovaya Elektron. (Moscow)* **25** 764 (1998) [*Quantum Electron.* **28** 744 (1998)]

[12]   Prokhorov A M, Tkachev A N, Yakovlenko S I *Dokl. Ross. Akad. Nauk* **329** 729 (1993) [*Phys.-Dokl.* **38** 158 (1993)]

[13]   Maiorov S A, Tkachev A N, Yakovlenko S I *Matem. Model.* **6** 13 (1994)

[14]   Mayorov S A, Tkachev A N, Yakovlenko S I *Laser Phys.* **4** 624 (1994)

[15]   Golyatina R I, Tkachev A N, Yakovlenko S I *Laser Phys.* **7** 449 (1997)

[16]   Golyatina R I, Tkachev A N, Yakovlenko S I *Kratk. Soobshsch. Fiz.* (1-2) 92 (1997)

[17]   Golyatina R I, Tkachev A N, Yakovlenko S I *Laser Phys.* **8** 1095 (1998)

[18]   Tkachev A N, Yakovlenko S I *Kvantovaya Elektron. (Moscow)* **20** 1117 (1993) [*Quantum Electron.* **23** 972 (1993]

[19]   Savel'ev V V, Yakovlenko S I *Kvantovaya Elektron. (Moscow)* **23** 1020 (1996) [*Quantum Electron.* **26** 984 (1996)]

[20]   Savel'ev V V, Yakovlenko S I *Kratk. Soobshch. Fiz.* (11 -12) 57 (1996)

[21]   Savel'ev V V, Yakovlenko S I *Laser Phys.* **7** 437 (1997)

[22]   Savel'ev V V, Yakovlenko S I *Kvantovaya Elektron. (Moscow)* **24** 939 (1997) [*Quantum Electron.* **27** 913 (1997)]

[23]   Golyatina R I, Syts'ko Yu I, Yakovlenko S I *Laser Phys.* **8** 860 (1998); *Kratk. Soobshch. Fiz.* (7) 3 (1998)

[24]   Letokhov V S, Mishin V I, Puretzky A A *Prog. Quantum Electron.* **5** 139 (1977)

[25]   Karlov N V, KrynetskiT B B, Mishin V A, Prokhorov A M *Usp. Fiz. Nauk* **127** 593 (1979) [*Sov. Phys. Usp.* **22** 220 (1979)]

[26]   Letokhov V S *Laser Photoionization Spectroscopy* (Orlando, FL: Academic Press, 1987)

[27]   Krynetskil B B, Mishin V A *Tr. Inst. Ohshch. Fiz. Akad. Nauk* **24** 59 (1990)

[28]   Greenland P T *Contemp. Physics* **31** 405 (1990)

[29]   Grossman M W, Shepp T A *IEEE Trans. Plasma Sci.* **19** 1114 (1991)

[30]   Meade T B, Schwartz M H *Nucl. Eng. Int.* **37** (460) 56 (1992)

[31]   Camarcat N, Lafon A, Perves J, Rosengard A, Sauzay G *Proc. SPIE Int. Soc. Opt. Eng.* **1859** 14 (1993)

[32]   Lorrain B, Sobrero R *Proc. SPIE Int. Soc. Opt. Eng.* **1859** 154 (1993)

[33]   Arisawa T, Maruyama Y, Suzuki Y, Kato M, Wakaida I, Akaoka K, Miyabe I, Ohzu A, Sugiyama A *Optoelectronics-Dev. Technol.* **8** (2) 203 (1993)

[34]   Morioka N *Proc. SPIE Int. Soc. Opt. Eng.* **1859** 2 (1993)

[35]   Guyot J, Gazalet J, Camarcat N, Figuet J *Trans. Am. Nucl. Soc.* **70** 777 (1994)

[36]   Cazalet J *Rev. Gen. Nucl.* (2) 20 (1996)

[37]   Coupin Y *Rev. Gen. Nucl.* (2) 24 (1996)

[38]   Scheibner C A, Stern R C, Worden E F *Trans. Am. Nucl. Soc.* **74** 130 (1996)

[39]   Mishin V A *Prikl. Fiz.* (2) 49 (1996)

[40]   Webb C E, in *High-Power Dye Lasers* (Berlin: Springer, 1997) p. 143

[41]   Aoki N, Kimura N, Konagai C, Shirayama S, Miyazawa T, Takahashi T *Proc. SPIE Int. Soc. Opt. Eng.* **1412** 2 (1991)

[42]   Dickinson G J *Nucl. Eng.* **33** 162 (1992)

[43]   Bass I L, Bonanno R E, Hackel R P, Hammond P R *Appl. Opt.* **31** 6993 (1992)

[44]   Hackel R P, Warner B E *Proc. SPIE Int. Soc. Opt. Eng.* **1859** 120 (1993)

[45]   Kostritsa S A, Mishin V A *Kvantovaya Elektron. (Moscow)* **21** 502 (1994) [*Quantum Electron.* **24** 464 (1994)]

[46]   Kostritsa S A, Mishin V A *Prib. Tekh. Eksp.* (5) 200 (1994)

[47]   Ko D K, Kirn S H, Kirn J B, Lee J M, Kostritsa S A, Mishin V A *Appl. Opt.* **34** 983 (1995)

[48]   Maeda M, Uchiumi M, Takao T, Oki Y, Igarashi K, Shimamoto K *Rev. Laser Eng.* **23** 752 (1995)

[49]   Igarashi K, Maeda M, Takao T, Uchiumi M, Oki Y, Shimamoto K *Jpn. J. Appl. Phys. 1* **34** 3093 (1995)

[50]   Kostritsa S A, Mishin V A *Kvantovaya Elektron. (Moscow)* **22** 542 (1995) [*Quantum Electron.* **25** 516 (1995)]

[51]   Grigor'ev I S, Mironov S M, Mikhailov I V, Preprint No. 5860/14 (Moscow: Institute of Atomic Energy, 1995)

[52] Watson J, Avicola K, Payne A, Peterson R L, Ward R *Proc. SPIE Int. Soc. Opt. Eng.* 2375 96 (1995)

[53] Vasil'ev S V, Mishin V A, Shavrova T V *Kvantovaya Elektron. (Moscow)* 24 131 (1997) *[Quantum Electron.* 27 126 (1997)]

[54] Vasil'ev S V, Kostritsa S A, Mishin V A *Zh. Tekh. Fiz.* 67 (3) 53 (1997) *[Tech. Phys.* 42 300 (1997)]

[55] Kostritsa S A, Mishin V A *J. Russ. Laser Res.* 17 322 (1996)

[56] Vasil'ev S V, Kuz'mina M A, Mishin V A *Dokl. II Vseros. Konf. Tiziko-Khimicheskie Protsessy pri Selektsii Atomov i Molekul¹ (Zvenigorod 1997)* (Paper presented at the Second All-Russia Conference on Physicochemical Processes in Selection of Atoms and Molecules, Zvenigorod, 1997) (Moscow: TsNIIatominform, 1997) p. 78

[57] Maximov 0 P, Mishin V A, Moshkunov A I *Proceedings of the International Conference Lasers 97, New Orleans, LA, 1997* (McLean, VA: STS Press, 1998)

[58] Vasil'ev S V *Kvantovaya Elektron. (Moscow)* 25 429 (1998) *[Quantum Electron.* 28 416 (1998)]

[59] Krynetskil B B, Kuz'mina M A, Mishin V A *Tr. Inst. Obshch. Fiz. Akad. Nauk* 24 50 (1990)

[60] Borisov S K, Mishin V A *Tr. Inst. Obshch. Fiz. Akad. Nauk* 24 3 (1990)

[61] ZhileTkin Ya M, KrynetskiT B B, Kuz'mina M A, Mishin V A *Zh. Tekh. Fiz.* 60 (5) 39 (1990) *[Tech. Phys.* 35 555 (1990)]

[62] Fomichev S V *J. Phys. D* 26 349 (1993)

[63] Kupecek P, Comte M, de Lamare J, Petit A *Proc. SPIE Int. Soc. Opt. Eng.* 1859 79 (1993)

[64] Wang Y 0, Baek I N, Suk C K, Young S K *New Phys. (Korean Phys. Soc.)* 33 31 (1993)

[65] Izawa Y, Nomaru K, Aly N E S, Кои K, Niki H, Nakai S, Chen Y-W, Yamanaka C *Proc. SPIE Int. Soc. Opt. Eng.* 1859 64 (1993)

[66] de Lamare J, Kupecek P, Comte M *Opt. Commun.* 95 305 (1993)

[67] Nomaru K, Izawa Y, Nakai S, Chen Y-W, Yamanaka C *J. Nucl. Sci. Technol.* 31 914 (1994)

[68] Chen Y-W, Yamanaka C, Niki H, Adachi S, Izawa Y, Nakai S *J. Opt. Soc. Am. B* 11 1585 (1994)

[69] Chen Y W, Izawa Y, Nakai S, Yamanaka C *Jpn. J. Appl. Phys. 1* 34 504 (1995)

[70] Borisov S K, Kuz'mina M A, Mishin V A *Kvantovaya Elektron. (Moscow)* 22 722 (1995) *[Quantum Electron.* 25 695 (1995)]

[71] Borisov S K, Mishin V A, Kuzmina M A *Priki Fiz.* (1) 65 (1995)

[72] Choe A S, Rhee Y, Lee J, Kuzmina M A, Misnin V A *J. Phys. B* 28 3805 (1995)

[73] Borisov S K, Kuz'mina M A, Mishin V A *J. Russ. Laser Res.* 17 332 (1996)

[74] Borisov S K, Kuz'mina M A, Mishin V A *Kvantovaya Elektron. (Moscow)* 25 177 (1998) *[Quantum Electron.* 28 169 (1998)]

[75] Borisov S K, Kuz'mina M A, Mishin V A *Dokl. II Vseros. Konf. 'Fiziko-Khimicheskie Protsessy pri Selektsii Atomov i Molekul' (Zvenigorod 1997)* (Paper presented at the Second All-Russia Conference on Physicochemical Processes in

Selection of Atoms and Molecules, Zvenigorod, 1997) (Moscow: TsNIIatominform, 1997) p. 53

[76] Yamada K, Tetsuka T, Deguchi Y *J. Appl. Phys.* **69** 6962 (1991)

[77] Ogura K, Arisawa T, Shibata T *Jpn. J. Appl. Phys. 1* **31** 1485 (1992)

[78] Doneddu F *Proc. SPIE Int. Soc. Opt. Eng.* **1859** 98 (1993)

[79] Watanabe J, Okano K *Phys. Fluids B* **5** 3092 (1993)

[80] Murakami M, Ueshima Y, Nishihara K *Jpn. J. Appl. Phys. 2* **32** L1471 (1993)

[81] Matsui T, Tsuchida K, Tsuda S, Suzuki K, Shoi T *Proceedings of the Sixth Latin-American Workshop on Plasma Physics, Sao Jose des Campos, Brazil, 1994* (Rio de Janeiro: INPE, 1994) Vol. 3, p. 313

[82] Demidova N S, Mishin V A *Zh. Tekh. Fiz.* **64** (7) 158 (1994) [*Tech. Phys.* **39** 719 (1994)]

[83] Nishio R, Yamada K, Suzuki K, Wakabayashi M *J. Nucl. Sci. Technol.* **32** 180 (1995)

[84] Demidova N S, Mishin V A *У. Moscow Phys. Soc.* **5** 223 (1996)

[85] Demidova N S, Mishin V A *Pis ma Zh. Tekh. Fiz.* **23** (8) 42 (1997) [*Tech. Phys. Lett.* **23** 311 (1997)]

[86] Demidova N S, Mishin V A *У. Moscow Phys. Soc.* **7** 201 (1997)

[87] Karlov N V, KrynetskiT B B, Kushlyanskil O A, Mishin V A, Nastyukha A I *7r. Fiz. Inst. Akad. Nauk SSSR* **114** 24 (1979)

[88] Bespalov 0 G, Karlov N V, Krynetskil B B, Kushlyanskil O A, Mishin V A, Nastyukha A I *Pis'ma Zh. Tekh. Fiz.* **3** 980 (1977) [*Sov. Tech. Phys. Lett.* **3** 402 (1977)]

[89] Nishimura A, Ohba H, Shibata T *J. Nucl. Sci. Technol.* **29** 1054 (1992)

[90] Magnaud J P, Claveau M, Coulon N, Yala P, Guilbaud D, Mejane A *Proc. SPIE Int. Opt. Soc. Eng.* **1859** 194 (1993)

[91] Berardo M, Marcellin J *Proc. SPIE Int. Soc. Opt. Eng.* **1859** 172 (1993)

[92] Anklam T M, Berzins L V, Hagans K G, Kamin G W, McClelland M A, Scarpetti R D, Shimer D W *Proc. SPIE Int. Soc. Opt. Eng.* **1859** 277 (1993)

[93] Nishio R, Suzuki K *Rev. Sci. lustrum.* **64** 3278 (1993)

[94] Nishio R, Suzuki K *J. Nucl. Sci. Technol.* **31** 572 (1994)

[95] Sytsko Yu I, Yakovlenko S I *Laser Phys.* **6** 989 (1996)

[96] Ishchenko V N, Karlov N V, Krynetskil B B, Lisitsin V N, Mishin V A, Razhev A M *Pis ma Zh. Tekh. Fiz.* **3** 1044 (1977) [*Sov. Tech. Phys. Lett.* **3** 428 (1977)]

[97] Ishchenko V N, Karlov N V, Krynetskil B B, Lisitsin V N, Mishin V A, Razhev A M *Tr. Fiz. Inst. Akad. Nauk SSSR* **114** 46 (1979)

[98] Karlov N A, KrynetskiT B B, Mishin V A, Stel'rnakh O M *Kratk. Soohshch. Fiz.* (6) 27 (1981)

[99] Borisov S K, Karlov N A, KrynetskiT B B, Mishin V A, Stel'rnakh O M, Prokhorov A M *Opt. Spektrosk.* **57** 780 (1984) [*Opt. Spectrosc.* (USSR) **57** 475 (1984)]

[100] Borisov S K, Kotochigova S A, Karlov N A, Mishin V A, Pentegov S Yu, Stel'rnakh 0 M *Opt. Spektrosk.* **61** 716 (1986) [*Opt. Spectrosc. (USSR)* **61** 448 (1986)]

[101]   Borisov S K, Karlov N A, Karulin F E, KrynetskiT B B, Mishin V A, Stel'rnakh 0 M, Pentegov S Yu *Opt. Spektrosk.* **62** 1216 (1987) *[Opt. Spectrosc. (USSR)* **62** 719 (1987)]

[102]   Borisov S K, Karulin F E, KrynetskiT B B, Mishin V A, Pentegov S Yu, Prokhorov A M, Stei'makh 0 M *Author's Certificate No. 1529510,* in *Byull. Izohret.* (46) 125 (1989)

[103]   Mishin V A, Pentegov S Yu *Opt. Spektrosk.* **69** 232 (1990) [*Opt. Spectrosc. (USSR)* 69 141 (1990)]

[104]   Yi J-H, Lee J-G, Kong H J *Phys. Rev. A* **51** 3053 (1995)

[105]   Yakovlenko S I *Usp. Fiz. Nauk* **136** 593 (1982) [*Sov. Phys. Usp.* **25** 216 (1982)]

[106]   Yakovlenko S I *Radiatsionno-Stolknovitel'nye Yavleniya* (Radiation-Collisional Phenomena) (Moscow: Energoatomizdat, 1982)

[107]   Gudzenko S I, Yakovlenko S I *Plazmennye Lazery* (Plasma Lasers) (Moscow: Atomizdat, 1978)

[108]   Derzhiev V I, Zhidkov A G, Yakovlenko S I *Izluchenie lonov v Neravnovesnoi Plotnoi Plazme* (Emission of Ions in a Nonequilibrium Dense Plasma) (Moscow: Energoatomizdat, 1986)

[109]   Anan'ev Yu A *Opticheskie Rezonatory i Lazernye Puchki* (Optical Cavities and Laser Beams) (Leningrad: Nauka, 1990)

[110]   Buchanov V V, Molodykh E I, Yurchenko N I *Kvantovaya Elektron. (Moscow)* **10** 1553 (1983) [*Sov. J. Quantum Electron.* **13** 1022 (1983)]

[111]   Boychenko A M, Zhidkov A G, Protopopov S V, Yakovlenko S I, Preprint No. 248 (Moscow: Institute of General Physics, Academy of Sciences of the USSR, 1987)

[112]   Zhidkov A G, Protopopov S V, Sereda 0 V, Terskikh A 0, Yakovlenko S I *Tr. Inst. Obshch. Fiz. Akad. Nauk* **21** 116 (1989)

[113]   Ramsey N F *Molecular Beams* (Oxford: Clarendon Press, 1956)

[114]   Ross J (Ed.) *Advances in Chemical Physics,* Vol. 10, *Molecular Beams* (New York: Wiley Interscience, 1966)

[115]   BelotserkovskiT 0 M *Chislennoe Modelirovanie v Mekhanike Sploshnykh Sred* (Numerical Simulation in Mechanics of Continuous Media) (Moscow: Fizmatlit, 1994) Chap. 4

[116]   Kogan M N *Dinamika Razrezhennogo Gaza* (Dynamics of Low-Density Gases) (Moscow: Nauka, 1967)

[117]   Bird G A *Molecular Gas Dynamics* (Oxford: Oxford University Press, 1976)

[118]   Mayorov S A, Tkachev A N, Yakovlenko S I *Izv. Vyssh. Uchebn. Zaved. Fiz.* **39** (1) 121 (1996)

[119]   GranovskiT V L *Elektricheskil Tok v Gaze* (Electric Currents in Gases) (Moscow-Leningrad: GITTL, 1952) Vol. 1

[120]   Kaptsov N A *Elektronika* (Electronics) (Moscow: GITTL, 1954)

[121]   Forrester A T *Large Ion Beams: Fundamentals of Generation and Propagation* (New York: Wiley, 1988)

Process Monitoring and Flow Control in Coaxial Laser Cladding

Jehnming Lin
Department of Mechanical Engineering
National Cheng Kung University
Tainan, Taiwan
Email: Linjem@mail.ncku.edu.tw
Fax: 886-6-2352973

ABSTRACT

A way of monitoring and controlling the clad profile and powder catchment in coaxial laser cladding is presented in this article. An estimation of the heat loss by conduction can be obtained from measurements of the surface temperature and performing an overall heat balance at the clad laser interaction zone. Through an inverse calculation of the boundary temperature based on the measurements, the powder catchment efficiency as well as the expected variation in the clad height during laser cladding can be estimated. The dynamic characteristics of the pneumatic system of a newly developed powder feeder were also investigated. However, only a brief introduction is made as rigorous analysis of the dynamic characteristics of the shroud gas system is beyond the scope of this study.

It was demonstrated that by altering the powder flow with various gas settings, a level and controlled clad profile could be achieved in multi-layer cladding. The results reveal a simple yet reliable way of monitoring and controlling the clad height and profile as required by near net shape manufacturing methods based on laser cladding.

1 Introduction

Lasers have been successfully used for surface cladding applications. The nature of the absorption of laser radiation, which is limited to the surface atoms for opaque materials, renders it an ideal surface heat source. Moreover, the ease of focus of the radiation meant that the heated zone can be accurately defined. This has enabled lasers to be developed as a precision cladding tool and by overlaying clad layers complex structures can be built. The only requirements are however to control and automate the process. Major problems encountered at present are the variations in the build height and the quality of the surface finish [1~3], which understandably boils down to the fundamental understanding of the interrelationship amongst all the process parameters.

Cladding temperature is one of the major processing parameters of laser cladding. Many papers [4~6] were dedicated to studying this parameter, but the works were mainly based on off-axis blown powder cladding. This differs from coaxial cladding, because the optical attenuation in the powder stream affects the laser energy absorbed on the clad surface. As mentioned in the referred paper [5], experimental results show that the laser attenuation in a coaxial powder stream might be as high as 50 %. Moreover some of this optical energy is not lost but arrives at the surface in the form of preheated powder.

There are many modeling studies of the cladding process. They include the heat flow, clad build, molten pool flow structures and other areas [4~6]. There have also been studies on the development of process monitoring by width sensors, plasma sensors [7] and acoustic sensors [8] as well as dilution sensors [9]. However there are few references to in-process cladding temperature measurement due to the powder interference on the cladding zone [10]. There appear to be none relating the temperature of the clad to the powder catchment efficiency or clad profiles as described here.

In this study, a least-square inverse approach is used to predict the heat absorption and by inference the clad height in laser cladding. The transient interface temperature was estimated by an inverse scheme, which only requires surface temperature measurements, taken by an IR pyrometer with fiber optics near the cladding surface. The heat conduction equation in the system was simplified to a 1-D cylindrical coordinate system, and the heat absorption was inversely solved by a finite element method. The variation of the clad height was predicted from the energy balance between the incoming powder stream and the irradiant power on the substrate. The results of the clad profile predicted from the measured temperature were compared with the clad profile by experiments. The powder feed was used as a control variable by adjusting the velocity of the shield gas. The dynamic characteristics of the pneumatic system of the powder feeder were investigated. By altering the powder flow with various gas settings, a level and controlled profile could be achieved.

2 Estimation of Catchment by Cladding Temperature

Due to the unknown energy absorption and the complexity of the powder catchment mechanism in the melt pool of laser cladding, Analytical solutions [11~13] for a moving point or line source can not be directly applied. It does, however, help at distances away from the melt pool.

Since the heat absorbed at the clad boundary is an unknown parameter, this type of problem can be solved as an inverse heat conduction problem [14]. Many inverse approaches have been adopted to solve the boundary heat flow conditions coupled with a few temperature measurements in some similar processes such as machining [15]. The cladding temperature problem can be treated as an inverse problem with in-process temperature monitoring. The scheme of the inverse estimation used for cladding catchment efficiency is explained as follows.

2.1 Energy Balance in Cladding

An energy balance on the control volume around the clad melt pool shown in Figure 1 gives a relationship between the heat delivered, that absorbed by the trapped powder and that lost by conduction, radiation and convection. In the coaxial powder feed system some of the powder arrives hot from having traveled down the beam. This heat absorbed by the powder, and therefore denied to the substrate, is handed back as the powder sticks to the clad and is melted into the clad layer. However, a small amount of hot powder will be lost. Melting powder is likely to stick so the lost powder will be the cooler fraction and therefore constitute a negligible heat loss. An energy balance on the clad melt pool is given in equation (1). It has assumed that most the powder arrives hot, there is no melting of the substrate and the thermal properties are constant.

$$Q = P(r)\beta(T) - h(T - T_a)A - \varepsilon\sigma(T^4 - T_a^4)A + \eta\dot{m}(H_p) \tag{1}$$

where:

Q	= heat lost by conduction from the pool [W];
$P(r)$	= incident laser power which varies with radial position, r [W];
$\beta(T)$	= surface absorptivity which varies with temperature T;
h	= heat convection coefficient $[Wm^{-2} K^{-1}]$;
A	= clad surface area $[m^2]$
T_a	= temperature of the surrounding air [K];
ε	= surface emissivity;
σ	= Stefan-Boltzmann constant of $5.6032 \times 10^{-8}[Wm^{-2} K^{-4}]$;
η	= powder catchment efficiency;
$\dot{m}$	= powder delivery rate $[kg\, s^{-1}]$;
H_P	= the enthalpy of the powder which includes the sensible and latent heat $[J\, kg^{-1}]$

The surface emissivity is affected by the beam absorption within the powder stream. Much of the energy absorbed by the powder is recycled into the melt but the powder will both shadow the substrate and scatter the incoming beam. The variation of heat loss by convection and radiation will be only a small amount compared to the optical energy absorbed by the powder during cladding and hence can be neglected. Thus assuming the laser beam is of steady power with constant thermal properties, any variation in the flux by thermal conduction will be due to variation in the powder absorption, and it is a function of the catchment efficiency. Hence:

$$\Delta Q = \Delta \eta m' H_P(T) \tag{2}$$

Equation (2) implies that the change of the heat flow into the substrate depends on the rate of powder catchment under stable cladding conditions.

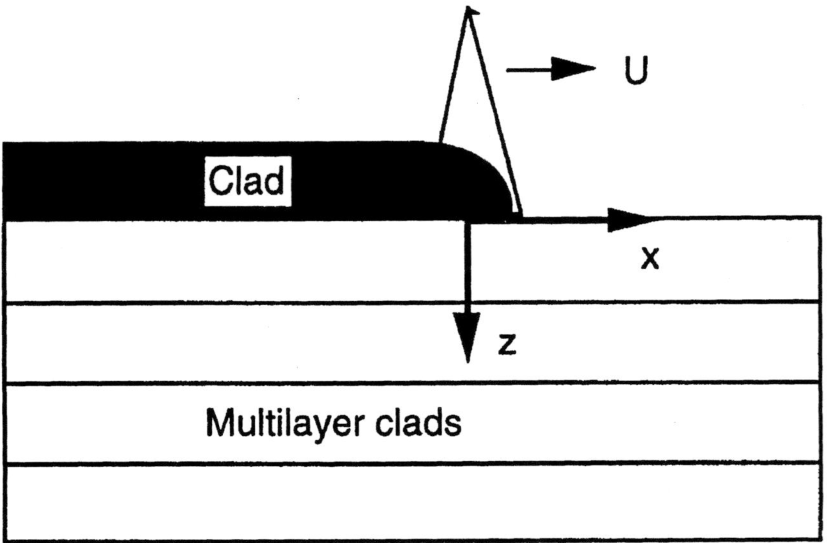

Figure 1: Coaxial laser cladding in multi-layer clad build-up.

2.2 Heat conduction on a thin clad wall

For the heat conducted in a thin wall as illustrated in Figure 2, in which the heat transfer in the y direction is neglected, it is assumed that no dimensional change of the top surface occurs after the clad build-up. The heat conduction equation of the thin wall with a moving heat source on the surface can be expressed as follows:

$$\frac{\partial^2 T}{\partial x^2} + \frac{\partial^2 T}{\partial z^2} + \frac{U}{\alpha}\frac{\partial T}{\partial x} = \frac{1}{\alpha}\frac{\partial T}{\partial t} \tag{3}$$

where α is the thermal diffusivity of the substrate, and U is the velocity of the substrate relative to the laser heat source.

Based on the assumptions of a line heat source across the wall edge and constant thermal properties of the substrate, no melting of the substrate and no heat loss from the side surface takes place. The solution for an unsteady-state problem involving a transportation term as given in equation (3) can usually be expressed as the product of an exponential term with a new dependent variable y. The symmetry of the temperature field allows it to be conveniently transformed to polar coordinates in the x-z plane. Hence, following the transformation suggested by Rosenthal [13], $T= \exp(-Ux/2\alpha)\psi\,(\,r\,,\,t\,) + T_0$ is substituted in equation (3) to yield the following differential equation.

$$\frac{\partial^2\psi}{\partial r^2} + \frac{1}{r}\frac{\partial\psi}{\partial r} - (\frac{U}{2\alpha})^2\psi = \frac{1}{\alpha}\frac{\partial\psi}{\partial t} \tag{4}$$

where $r = \sqrt{x^2 + z^2}$, with the boundary and initial conditions

$$\lim_{r\to 0}(-2\pi r g k\,\frac{\partial\psi}{\partial r}) = Q \tag{5a}$$

$$\psi_{r\to\infty} = 0 \tag{5b}$$

$$\psi(r,0) = 0 \tag{5c}$$

In the literature the exact solution for steady-state heat flow problem in a thin plate is expressed as equation (6) [13]. This can be applied to solve the temperature field of a 2D cladding thin wall.

$$\frac{2\pi(T - T_0)kg}{Q} = \exp(\frac{-Ux}{2\alpha})K_0(\frac{Ur}{2\alpha}) \tag{6}$$

where k is the thermal conductivity of the substrate, g is the thickness of the thin plate, Q is the heat flow of the line heat source lying on surface in y direction and T_0 is the substrate temperature at infinity, K_0 is the modified Bessel function of the second kind and zero order.

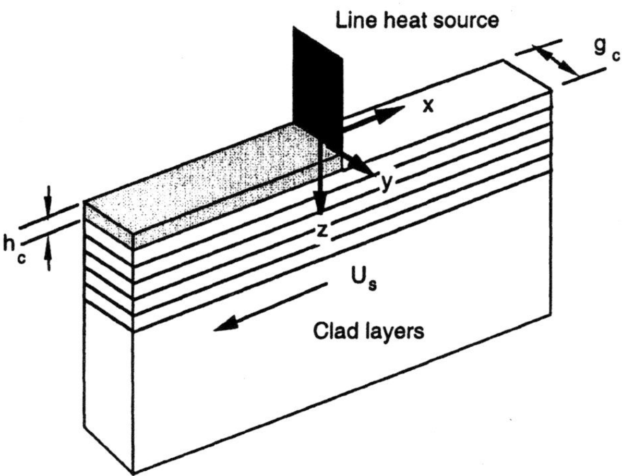

Figure 2: Configuration of the heat conduction problem for coaxial laser cladding.

2.3 Inverse method for solving heat conduction problem

If the heat flux or temperature histories at the surface of a solid are known functions of time, then the temperature distribution can be solved. This is termed a direct problem. Otherwise, the prediction of the unknown interface temperature of a solid by the temperature measured in the object would be treated as an inverse heat conduction problem. In many dynamic heat transfer situations, the surface heat flux and temperature histories of a solid might be determined from transient temperature measurements at one or more interior locations. In fact, during the past two decades, this problem has become important [14]. Many methods have been proposed to solve the inverse heat conduction problem either analytically or numerically. Here a least-square inverse scheme with a finite element method [14,16] is introduced and applied to solve the unknown heat flux in coaxial laser cladding. This is shown in Figure 3 for a one-dimensional heat conduction system with unknown boundary conditions of heat flux $q(\xi_0,t)$ on the left boundary and constant temperature To in the infinite domain of the system, where $Y(\xi_j,t)$ is the measured temperature at an interior point ξ_j. The inverse method developed by Beck [14] to solve a 1D heat conduction problem is presented here. In Beck's approach, he focused on the observation that the interior temperature response is delayed and damped with respect to the change of the boundary condition of the system. As shown in Figure 3, the surface heat flux $q(\xi_0,t)$ at time t can be determined by using interior temperatures $Y(\xi_j,t)$ measured at a time larger than t. These temperatures are so called "future temperatures"[14].

The objective of this inverse method is to select q_m to achieve the closest agreement in a least square sense between the computed and measured temperatures at position ξ_j in a specified time interval, and it can be expressed as

$$f(q_m) = \sum_{j=1l}^{ml}\sum_{i=1}^{mr}(Y_{j,m+i-1} - T_{j,m+i-1})^2 \tag{7}$$

where Y, T are the measured and estimated temperatures respectively; i,j are the index of the time step and interior sensors; ml is the number of the interior points; mr is the number of the future temperatures.

By minimizing the function $f(q_m)$, the value of q_m corresponding to the minimum of $f(q_m)$ can be found. This is accomplished as $\partial f/\partial q_m$ is equal to zero or

$$2\sum_{j=1l}^{ml}\sum_{i=1}^{mr}(Y_{j,m+i-1} - T_{j,m+i-1})(\frac{\partial T_{j,m+i-1}}{\partial q_m}) = 0 \tag{8}$$

Since $T_{j,m+i-1}$ and $\partial T_{j,m+i-1}/\partial q_m$ in equation (8) are functions of heat flux and can be evaluated at the minimum of $f(q_m)$. A Taylor series expansion of $T_{j,m+i-1}$ is taken at a time step of t_{m-1} for a change of heat flux $\Delta q = (q_m - q_{m-1})$. Neglecting higher order terms, it gives

$$T_{j,m+i-1} = T^*_{j,m+i-1} + (q_m - q_{m-1})T^*_{j,m+i-1;q} \tag{9}$$

Here the notation $T^*_{j,m+i-1}$ (i=1~mr) means temperature evaluated by using heat flux q_{m-1} at time t_{m-1} and the term

$$T^*_{j,m+i-1;q} = \frac{\partial T_{j,m+i-1}}{\partial q_m} \tag{10}$$

is the so called sensitivity coefficient and calculated at time t_{m-1} [14].

Substituting equation (9) into (8), the heat flux q_m at the minimum of $f(q_m)$ can be solved as

$$q_m = q_{m-1} + \frac{1}{\Delta M}\sum_{j=1l}^{ml}\sum_{i=1}^{mr}(Y_{j,m+i-1} - T^*_{j,m+i-1})T^*_{j,m+i-1;q} \tag{11}$$

where

$$\Delta M = \sum_{j=1l}^{ml}\sum_{i=1}^{mr}(T^*_{j,m+i-1;q})^2 \tag{12}$$

For a linear case, the sensitivity coefficients in equation (11) are invariant and do not need to be recalculated for each value of q_m. Beck also provided a simple method to calculate the sensitivity coefficients as follows :

Replace q in equation (5a) by q_m and the initial time by t_{m-1}. Let the sensitivity coefficient be denoted as

$$T_q = \frac{\partial T^*}{\partial q_m} \tag{13}$$

Taking the derivatives of equation (3) with respect to q_m , the linear sensitivity equations will yield exactly the same form to the original heat conduction problem. This means that the computation of temperature T or ψ can be utilized in determining the sensitivity coefficient T_q [14].

A finite element program with least-square inverse scheme was developed, and a flow chart of this computer program is shown in Table 1, which is based on equations (7) to (13).

From equations (2), (3), (10) and (11), the variation of the powder catchment due to the change of the input energy on the thin plate edge can be inversely calculated if the temperature within the substrate is known for the cladding process.

Table 1: Flow chart of the inverse estimation.

1. Read the data file of the temperature-time from experimental results
2. Assume initial values of q_m to calculate the sensitivity coefficient T_q
3. Calculate and replace q_m by the least-square inverse method
4. Calculate the transient temperature field
5. Check the convergence of the temperature computation with the temperature data from experiment. If the result is not convergent, go to step 3
6. End

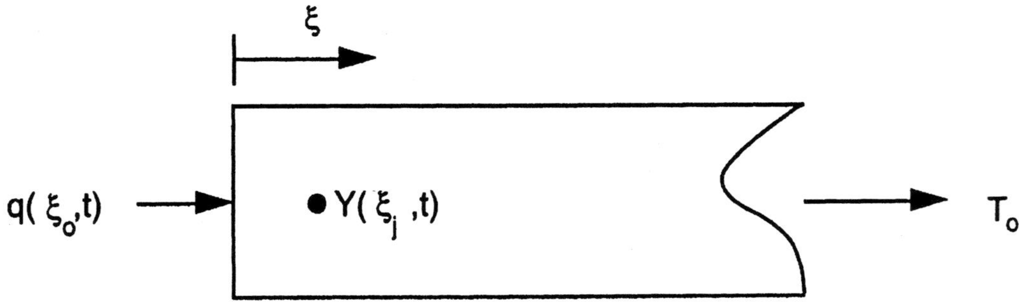

Figure 3: Inverse problem.

3 Clad Temperature Monitoring

3.1 Experiment arrangement

The arrangement for detecting the temperature at a distance beneath the clad track is shown in Figure 4. A thin vertical plate was built by multiple cladding technique using the coaxial nozzle. The temperature was detected with an IR pyrometer and fiber optic. The experiments took the form of cladding under various powder catchment conditions by varying the shield gas flow under a 1 kW CO_2 laser radiation. The temperature signals were sampled at a rate of 1 kHz and stored on a personal computer for further analysis.

3.2 IR pyrometer

The pyrometer (THERMO-HUNTER) used here is based on an IR photo detector (Ge) with a spectral sensitivity in the wavelength of 1.0 to 1.6 μm. The temperature range of the system is rated as 400 to 1000 ^{0}C. This device could be used for directly viewing through an infrared glass fiber and centering the sensitive area on the required location. The signal output was linearlized by a pre-amplifier installed inside the pyrometer. The pyrometer system has a measuring response time of 0.05 sec and an accuracy of 1.5 ^{0}C in the calibration of the black-body radiation.

The fundamentals of the IR photo detector in temperature measurement have been fully discussed in the referred paper [17]. There is problem of a steep thermal gradient on a large viewing area using this instrument in laser cladding. To overcome this drawback, a optical fiber was used to reduce the viewing area. The fiber was made of zirconium fluoride with a diameter of 0.24 mm and optical aperture of 12°. The average transmissivity of this type of fiber is about 72% (fiber length of 1 m) in the infrared wavelengths.

In order to further improve the measurement accuracy of the pyrometer, a stainless steel tube with a length of 5 mm and inside diameter of 0.2 mm was used to reduce the viewing angle to 2.3^{o} of the optical fiber and avoid the interference of the powder streams. To compensate for the emissivity changes of the substrate materials during cladding, a performance test was made by viewing the optical fiber on the substrate materials heated in an electrical furnace with various temperatures. The response curve of the pyrometer system is illustrated in Figure 5.

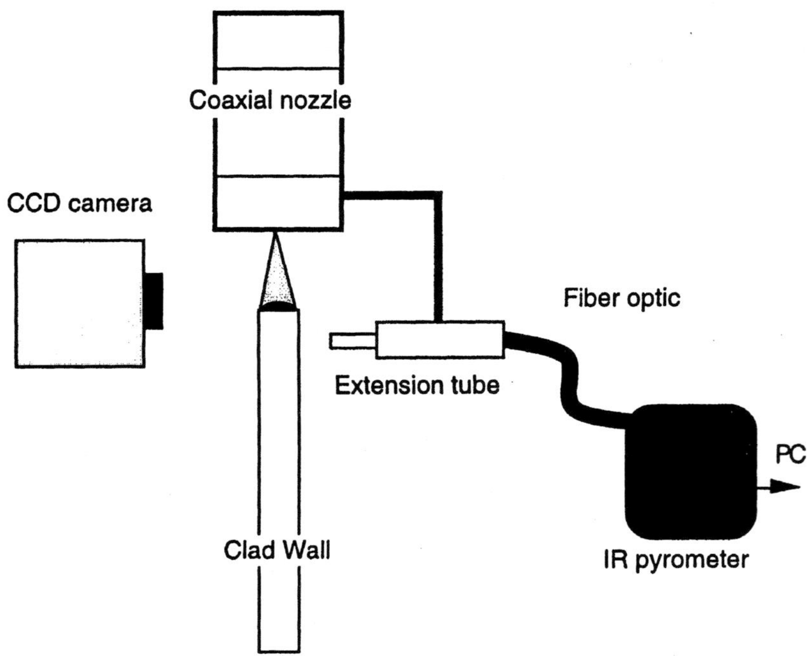

Figure 4: Cladding temperature and profile monitoring (front view).

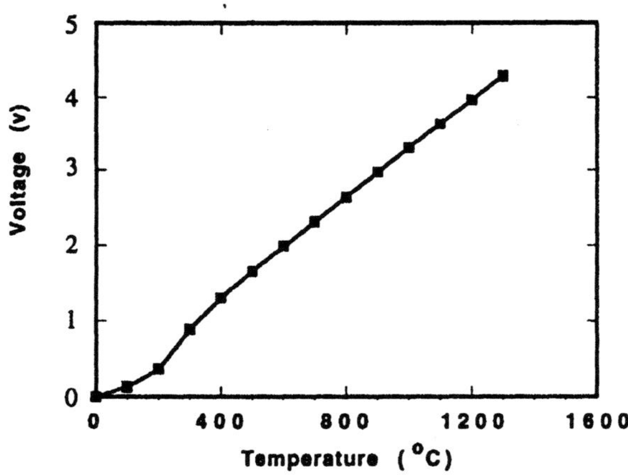

Figure 5: Calibration data of the mild steel substrate in an electrical furnace heating.

4 Experiment

Cladding experiment was carried out with various shield gas settings at constant powder mass flow rate and substrate speed to generate various thickness of the clad layers under 1 kW CO_2 laser radiation. The travelling path of the laser beam was arranged as illustrated in Figure 6, where the laser was turned on in front of the wall edge and turned off after the wall end for each single pass of the cladding track. Simultaneously the surface temperature under the cladding area was measured by the IR pyrometer through a fiber optic. The fiber optic was mounted onto to the coaxial nozzle head to give a fixed viewing position beneath the cladding zone.

A mild steel plate of 2x50x100 mm^3 was used as a substrate and coated on its longitudinal edge with 304L stainless steel powder with an average particle diameter of 75 μm to simulate a wall build. The substrate speed was set at 5 mm/s. An effective clad width about 2 mm along the straight edge of the substrate at a stand off distance of approximately 8 mm from a coaxial cladding nozzle was arranged. In order to measure the temperature near the clad surface under the laser radiation, the fiber optic was implemented to follow the laser beam and view the clad layer at a depth of 0.5 mm beneath the cladding interface with a viewing diameter of 0.3 mm in the sensor location. The cladding condition was selected at an average powder feed rate of 0.05 g/sec with a flow velocity of argon gas U_c of 4 mm/s to carry the powder. Various velocity settings of the outer shield argon gas U_i from 0 to 6 m/s and inner shield argon gas U_o from 2 to 4 m/s in an inward nozzle were arranged. The room temperature was 22 ^{0}C during the experiments.

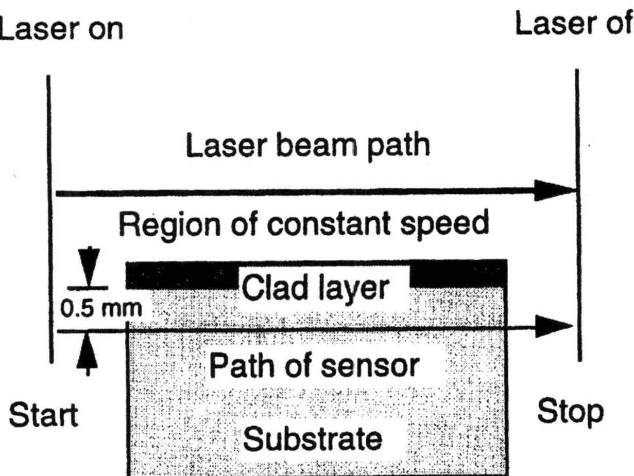

Figure 6: Traveling path of the laser beam in the wall edge cladding test.

4.1 Results of Temperature Measurement

The catchment data for three testing conditions are shown in Table 2. The clad surface temperature was measured for various flow conditions as shown in Figure 7. The quasi-steady state of the measured temperature at the leading edge was achieved less than 1 sec after the cladding process.

The results of the measured temperature compared with the edge profile at the finish end are illustrated in Figure 8. It can be seen that the enlarged edge profiles closely correspond to the rise in the temperature data.

Table 2: Thermal properties of the substrate and powder materials.

Material	Thermal diffusivity α (m^2 s^{-1})	Volume thermal capacity ρC_s (J m^{-3}K^{-1})	Thermal conductivity k (J m^{-1}s^{-1}K^{-1})	Fusion heat H_f (Jkg^{-1})
Carbon steel	9.1x10^{-6}	4.5x10^6	41.0	2.5x10^5
Stainless steel	(304L) 5.3x10^{-6}	4.7x10^6	24.9	2.7x10^5

Table 3: Powder catchment of three testing cases.

Test case [U_0,U_1] (m/s)	Powder delivery (g)	Powder catchment(g)	Catchment efficiency (%)
a [2,0]	0.92	0.30	33
b [4,0]	1.04	0.28	27
c [4,6]	0.90	0.21	23

4.2 The Relationships between the Clad Profile Generation and Measured Temperatures

It is shown in Figure 8 that the change of the substrate temperature will significantly affect the powder catchment and the generated clad profiles. It can be noticed that there is a temperature jump that correlates to the clad ramp at the end edge during each cladding test. Similar results have also been noticed at the starting edge. However, the heat penetration at the finish end is much more significant due to the higher substrate temperature from a longer interaction time with the laser beam. An experimental verification with the inverse estimation of the surface energy variation by the clad temperature measurement is made in the following section. A relationship between the inverse prediction and the clad profile measurements is also established.

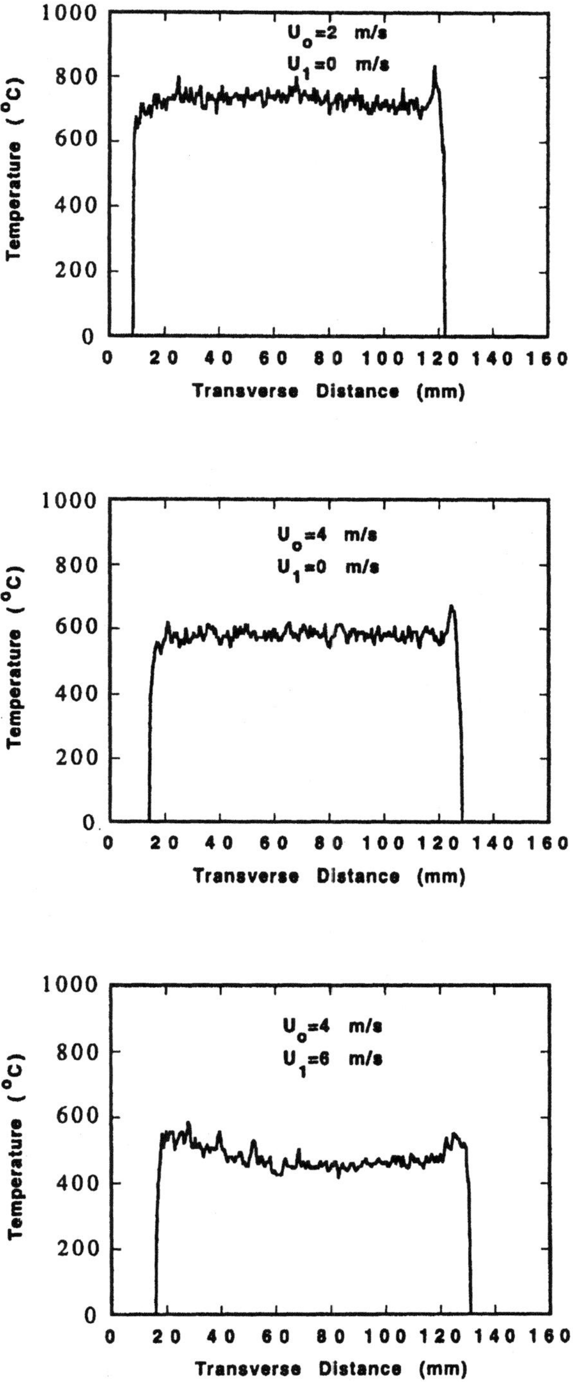

Figure 7: Measured results of the clad temperature under various shield gas settings.

                    Jehnming Lin

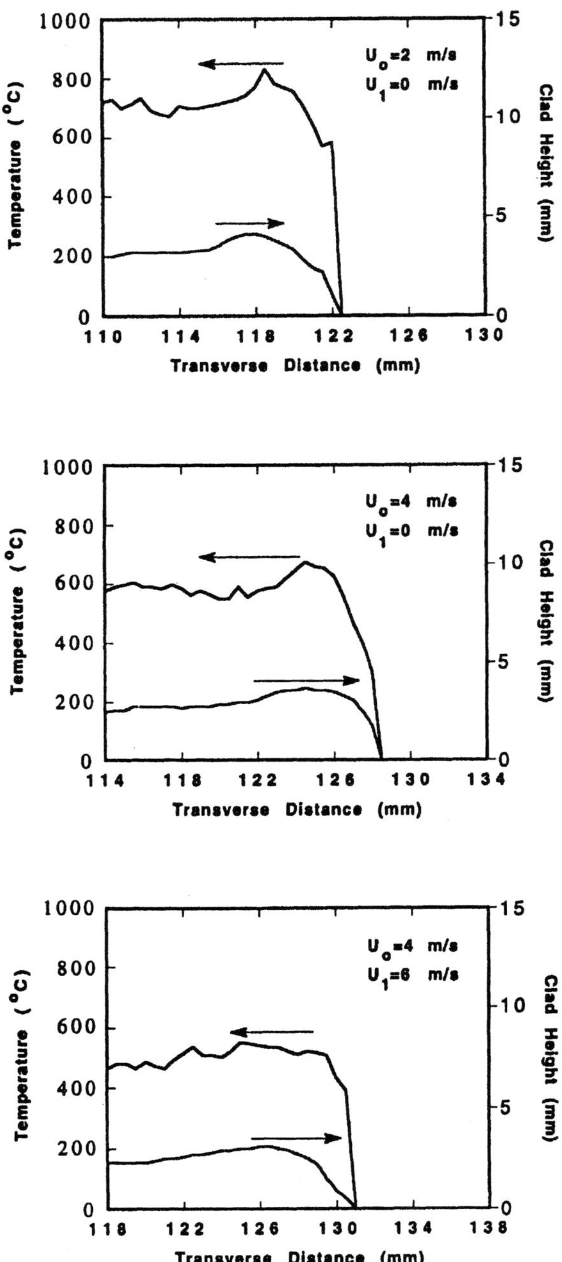

Figure 8: Clad temperature and profile at the end edge of the cladding test under various shield gas settings.

4.3 Results of Inverse Estimation on the Powder Catchment

Based on the experimental results discussed previously, the unsteady state solution of the heat input at the cladding area was solved by the inverse method. From the measured temperatures in Figure 7, the corresponding total heat input on the cladding surface was calculated and plotted in Figure 9.

The inverse estimated results of the heat flow by conduction, Q, were calculated at future temperatures of mr=2 and a time interval of δt=0.5 sec. From equation (9), the ratio of the deviation of the catchment efficiency to the average catchment efficiency weighted in the experiment is plotted in Figure 10, against the ratio of profile deviation at the track end to the average clad height under different gas flow conditions. It shows a near linear relationship with a slope of 4.7. The non-uniform thermal loading at the open end is one of the possible explanations for the edge height change. The open edge boundary with a higher thermal resistance will retard the heat flow and induce a larger quantity of heat to accumulate there. A similar defect has also been noticed at the starting edge.

5 Clad Profile Monitoring and Control

The clad build-up in the cladding process was monitored by an IR pyrometer and a CCD camera as illustrated in Figure 4. A valve switch was used for the control of the powder flow and catchment to compensate for any profile error.

5.1 Dynamic Characteristics of the Shroud Gas System on the Powder Catchment Control

The valve switch used here is illustrated in Figure 11. It is a two-way on/off gas solenoid valve, which can withstand a maximum gas pressure of 10 bar. The DC signal from the switching of the valve switch was monitored for dynamic characterizations. It is shown in Figure 12 that the electrical response time of shutting on/off this switch is on average less than 1 ms.

For measuring the dynamic response of the gas flow in the pipe and the nozzle, a solenoid valve was connected to the gas pipe system of the coaxial nozzle. An IR pyrometer system was adopted with an objective lens (diameter of 8 mm and a focal length of 17 mm) to detect radiation from the hot powder stream. The measured radiation for a gas pressure of 0.5 bar is shown in Figure 13. Since the time lag of the pipe system is much larger than the response time of the pyrometer system and the melting time (about 0.01 s) [17] of the particles in the coaxial gas stream, the measured radiation can be used to analyze the response time of the powder stream in coaxial laser cladding.

                                                           Jehnming Lin

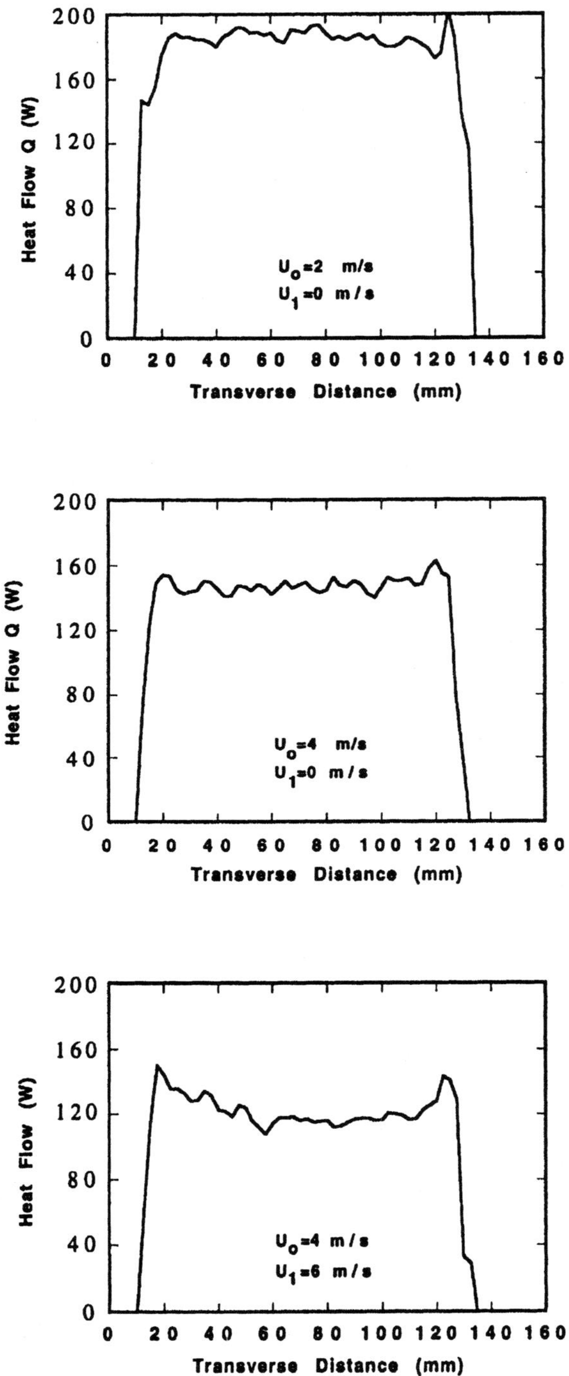

Figure 9: Estimated results of the heat flow on the clad surface by inverse method.

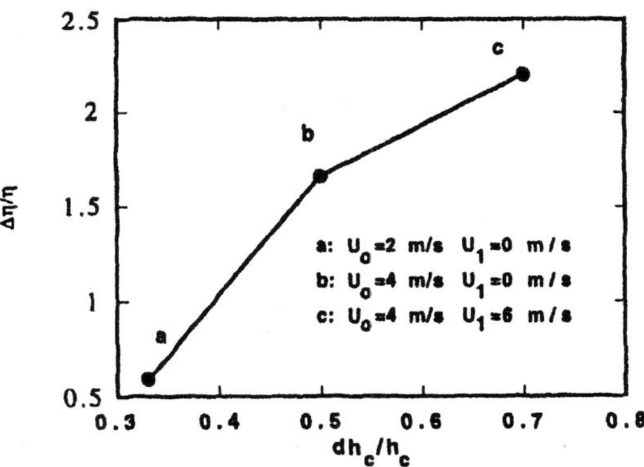

Figure 10: Relationship between the change of the catchment efficiency and the clad height at the wall end.

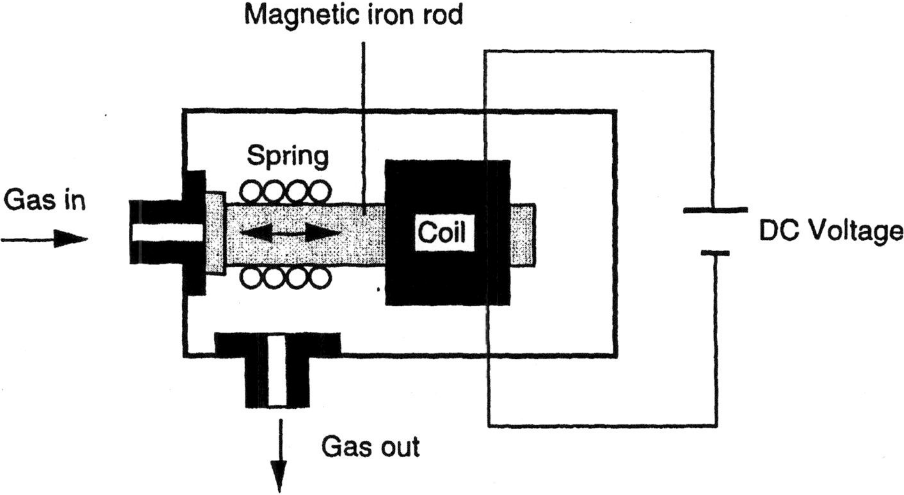

Figure 11: Structure of the 2 way pneumatic value.

The time delay corresponding to the input pressure during pressure rise and pressure drop was measured separately. It was found that the time lag for the former is shorter than the latter. The difference in the time lags could possibly be attributed to different stroke accelerations or forces in the solenoid valve at the open/close position and also to line capacity effects. On average, the time lag of the 2 way valve is about 0.15 sec for a 3 m pipe. As illustrated in Figure 12, when the valve is switched off standing pressures in the inlet pipe will not only resist the closure of the valve, but also force it to open. A spring-type pressure regulator has been used to damp or regulate the flow pressure to an even stroke condition, but it results in a long response time of the flow pressure in the pipe line [18].

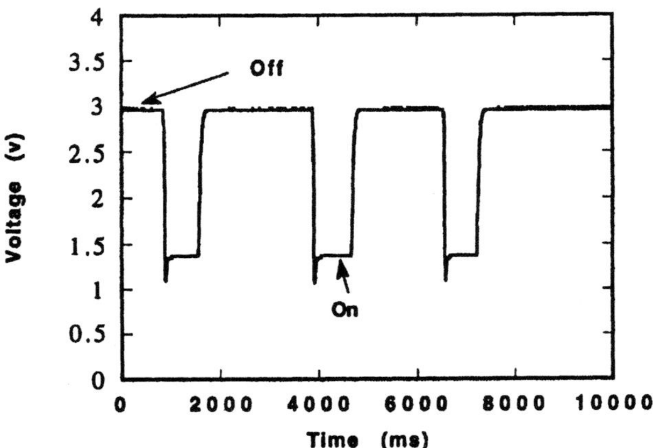

Figure 12: Electrical signal of the valve switching.

An understanding of the dynamic behavior is required for a proper design of the pneumatic system. However, a rigorous analysis of the dynamic characteristics of the shroud gas system is beyond the scope of this study, hence only a brief introduction is made on the characteristics of the pneumatic system and on the dynamic testing. The influence on the dynamic response of a flow to valve adjustment in an elastic pipe with a uniform cross-section is discussed below. The results can be applied to the coaxial nozzle design for the purpose of system control.

Normally, the analysis of pneumatic lines aims at two phenomena: that a finite time is taken for a pressure signal to traverse the length of the line, and that a standing wave can occur in the line. The result of the first phenomenon is that a pressure pulse at one end of the line does not cause the pressure at the other end to begin to rise instantly. The result of the second phenomenon is that the phase lag for harmonic inputs exceeds 180 degrees and passes the natural frequency of the column of gas [18]. It is apparent that pressure variations along the line must be considered to predict the effects of these phenomena. A simplified formulation is considered adequate for most conventional engineering applications and is available in Reference [18].

The general equations of the transient pressure and velocity in the pipe line are derived from the ideal gas equation, the momentum equation and the continuity equation with the assumption of a fully developed flow field [18]. Equation (17) is a first order solution of 1D flow equations for a long pipe-line with damping ratio not equal to unity. This solution is simple but adequate in most qualitative analysis [18]. It is adopted in most works on pipe-line design for quick reference of the time lag of a system.

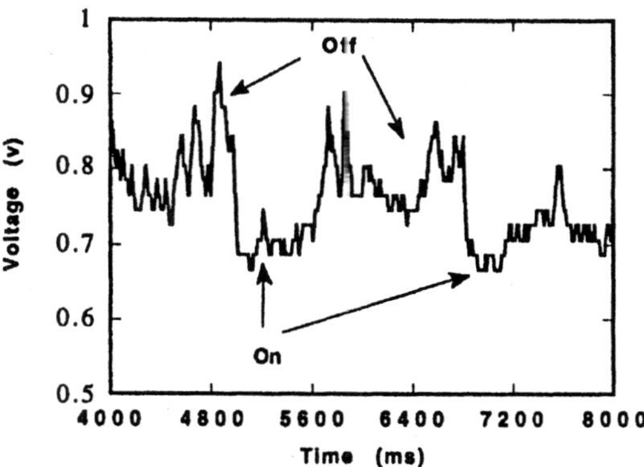

Figure 13: Signal from IR pyrometer for a hot stream with the valve switching on/off.

$$\frac{p_2}{p_1} \approx 1 - \exp(-\frac{t}{\tau_L}) \tag{17}$$

The first-order time constant τ_L is , which is the time lag for the outlet pressure to reach 63% of the inlet pressure in a pipe system.

$$\tau_L = \frac{\mu L(\frac{1}{2} AL + V)}{n_s p D^2 A} \tag{18}$$

where V is the volume of the end chamber, L is the length of the pipe line, D and A are the inside diameter and the cross-sectional area of the pipe respectively, n_s is a constant between 1 (for constant temperature) and the ratio of the specific heat (for constant entropy), p_1, p_2 and p are the inlet, outlet and average gas pressure respectively and μ is the flow viscosity.

Neglecting the effect of the end chamber volume, V, of the coaxial nozzle and assuming a constant temperature, Equation (18) can be rearranged as

$$\tau_L = \frac{\mu}{2p}(\frac{L}{D})^2 \tag{19}$$

It is seen that the time constant τ_L is a function of the aspect ratio of the pipe, L/D, and is inversely proportional to the average pressure, p. It means that the time constant will decrease with reducing aspect ratio of the pipe and/or with increasing input pressure.

Therefore a quick flow response can be obtained by reducing the pipe length and/or increasing either the pipe diameter or input pressure.

5.2 Results of Clad Profile Control

The feasibility of varying powder catchment to eliminate if not minimize profile distortion in coaxial cladding was investigated. The in process control was achieved by on-line profile monitoring and manual adjustment of the powder catchment.

A stainless steel cladding sample 30 mm long and 2 mm wide with 30 layers of straight clad was made in the test. The process condition was the same as the cladding case in Section 4.1. Based on the profile variation, which was estimated from the first clad layer through on-line temperature measurement, an adjustment of the powder catchment was carried out by varying the shield gas velocity in the coaxial nozzle. This was to compensate the profile distortion at the wall edges leading to the next clad layer. The amount of compensation was based on the catchment chart with various stream settings as shown in Table 2. Assuming an equal amount of distortion at each layer, the edge profile, which was monitored with a CCD camera, can thus be improved.

Constant powder delivery was applied at each layer in both the forward and backward directions. In-process control of the clad profile was achieved by switching the gas valve to correspond with the dynamic characteristic of the valve switch in the flow system as discussed in the previous section. The response signal of the valve system was measured and shown in Figure 13.

Figure 14 shows a section of the stainless steel wall which has a mid-clad height of 3.5 mm. The deposition on the right edge was controlled by the valve switch while that on the left edge was made without any control, thus resulting in a peak height of 4.2 mm. With proper temperature measurement and profile monitoring, the edge ramp on the right end could be reduced to some extent by adjusting the gas flow rate. However as shown in Figure 14, a rounded edge still existed. This is because the jet stream will cause an uneven powder distribution at the vertical edge. Moreover, the high temperature occurring at the open edge will re-melt the edge and aggravate this problem.

6 Conclusions

The heat conduction model adopted in this study was transformed to an inverse problem. The least square inverse method can be extensively used to solve unsteady state and non-linear heat conduction problems resulting from the change of thermal properties in processes. Heat conduction from the cladding area was mainly influenced by powder deposition and substrate geometry. The energy carried by the powder stream forming the clad on the substrate was dissipated as heat in the cladding process and this contributes significantly to the non-linearity of the edges in multi-layer clad generation.

(a)

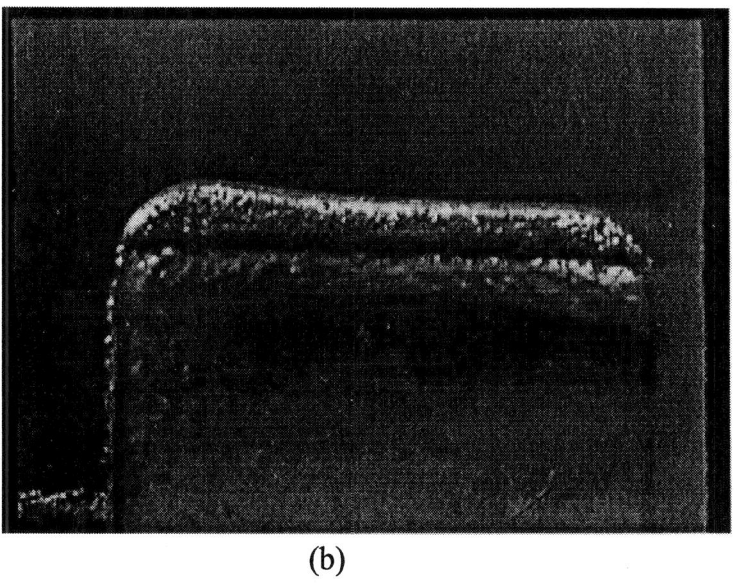

(b)

Figure 14: Profile control on multi-layer cladding, (a) in process, (b) final result.

The inverse method has successfully predicted the heat flow in laser cladding through clad temperature measurement. The change of the heat flow on the clad surface is related to the variation of powder catchment efficiency, which also affects the edge profile in multi-layer clad walls. This method is a potential tool for monitoring clad catchment in flat clad profile applications such as rapid prototyping by weld-build processes.

The predicted catchment variations due to uneven temperature profiles of the thin wall substrate were confirmed by experiment. A linear relationship was established between the clad profile and the heat flow as estimated by the inverse method. It was shown that the variation of the heat flow on the substrate could affect the powder catchment, causing possible thermal distortion at an open wall edge.

It was also shown that varying the powder concentration, which can be achieved through controlling the flow rate of the shroud gas to vary the catchment efficiency, could eliminate the problem of profile distortion for open wall edges in laser cladding. This approach appears to be capable of monitoring powder catchment efficiency and hence the clad-build in spite of the uncertainties of thermal properties at high temperatures and the non-uniform temperature field observed by the optical sensor. However the inverse method provides an insight into the cladding process in terms of the control of the powder catchment and the clad build.

Acknowledgement

Many thanks are due to Professor W. M. Steen of the University of Liverpool for his invaluable suggestions to this work, which was funded by NSC, ROC and carried out as part of the LEMA (Laser Engineering for Manufacturing Applications) project of the EPSRC, UK.

References

[1]   Koch, J. L. and Mazumder, J., " Rapid Prototyping by Laser Cladding ", ICALEO (1993) 556-559.

[2]   Ashley, S. " Rapid Prototyping System " Mechanical Engineering (1991)113, 34-43.

[3]   Murphy, M., Lee, C. and Steen, W. M. " Studies In Rapid Prototyping By Laser Surface Cladding " , ICALEO (1993) 882-891.

[4]   Weerasinghe, V. M., *Laser Cladding of Flat Plates*, PhD Dissertation, Imperial College of Science and Technology (1984)

[5]   Vetter, P.-A. Engel, Th. and Fontaine, J. " Laser Cladding: The Relative Parameters for Process Control ", SPIE (1994) 2207, 452-462.

[6]   Hoadley, A. F. A. and Rappaz, M. " A Thermal Model of Laser Cladding by Powder Injection" Metallurgical Trans. B. (1992) 23, 631-642.

[7]   Li, L., Brookfield, D. J. and Steen, W. M. " Plasma Charge Sensor for In-Process, Non-contact Monitoring of The Laser Welding Process ", Meas. Sci. Technol. (1996), 7, 615-626.

[8]   Jon, M. C. " Non-contact Acoustic Emission Monitoring of Laser Beam Welding ", Welding J. (1985) 64, 43-48.

[9]   Li, L and Steen, W. M. " A Dual-Frequency Electromagnetic Sensor for Non-contact Dilution Evaluation in Laser Cladding and Alloying Processes ", Meas. Sci. Technol. (1996) 7, 650-660.

[10] Li, W. -B, Powell, J. and et al. " Redistribution of The Beam Power in Laser Cladding by Powder Injection ", Laser in Engineering, (1996) 5, 175-183.

[11] Steen, W. M. , *Laser Material Processing*, Springer-Verlag (1991).

[12] Cline, H. E. and Anthoney, T. R. " Heat Treating and Melting Material with A Scanning Laser or Electron Beam ", J. of Applied Physics (1977) 48, 3895-3900.

[13] Carslaw, H. S. and Jaeger, J. C. *Conduction of Heat in Solids*, Oxford (1959).

[14] Beck, J. V. , B. Blackwell and C. R. St. Clair, *Inverse Heat Conduction: Ill-posed Problems*, Wiley (1985).

[15] Ohadi, M. M. and Cheng, K. L. " Modeling of Temperature Distributions in The Workpiece During Abrasive Water Jet Machining ", ASME Trans. J. of Heat Transfer (1993) 115, 447-452.

[16] Becker, E. B. and etc. *Finite Elements An Introduction*, Prentice - Hall (1981)

[17] Lin, J. M. " Temperature Analysis of The Powder Streams in Coaxial Laser Cladding ", Optics and Laser Technology, Vol. 31, pp. 565-570, Nov. 1999.

[18] Andersen, B. W. *The Analysis and Design of Pneumatic Systems*, Robert E. Krieger Pub. Co. (1976).

Lasers with Fast Electronic Synthesis of Spectral Line Shape

Parkhomenko Yu. N. [1], Anisimova O. V. [1], Galkin O. N. [2]
[1]Institute for Sorption and Endoecology Problems of National Ukrainian
Academy of Sciences
[2]Department of Chemical Engineering, University of Houston
4800 Calhoun Rd, Houston, TX 77204-4004

We have developed principles for fast electronic tuning of a spectral line shape in short-pulse lasers on the base of the separate control of phase and amplitude transmittance of dispersive cavity. We study formation of the spectral function of the diffractively coupled cavities and of that with its given moments. The influence on spectral kinetics of nonlinear processes in the active medium is investigated in cases of a single spectral line and of a composite spectrum. The algorithm for tuning basic spectral moments is substantiated. The continuous spectra synthesis is realized and investigated experimentally in the laser with a novel control device including a spatial acousto-optic modulator and a lens telescope.

1 Introduction

The development of effective media possessing power wide gain spectral lines (dyes, media with electron-phonon transitions: corundum with titan ion impurity, alexandtite, forsterite etc.) and methods of quick electronic tuning [1] has defined opportunities of tunable lasers and their essential role in many scientific and technical applications recently. Most applications are based on narrow spectrum tuning over a wide range. The up-to-day systems allow for the elaboration of lasers with quick synthesis and tuning of the given complex spectrum (discrete multiline or analog continuous one). These lasers will possess a great potential for the development of new, more effective methods in a variety of applications: 1) quick spectroscopy by use of complex laser spectra that correlate to absorption spectra of substances under study (these techniques allow also to increase the signal-to-noise ratio); 2) quick formation and switching of the complex directional diagram of the laser emission transformed in the device with the angular dispersion; 3) the optimum influence on systems with complex spectra: selective or resonance actions, information records, and so on. The practical value of such laser development is also supported by the increased efficiency in solving of several problems in communication [2,3,4], fast processes analysis [5,6,7], and lidar technique [8-11] if they use multifrequency spectra.

In this paper we will further develop the problem of analog laser spectrum formation, which was first posed by us in [12] and is important for the above-mentioned applications 2, 3 and for other ones. We will consider the laser with an acoustooptic device because it is the most effective and convenient unit for intracavity application.

In our previous papers the formation (synthesis) of discrete spectra (consisting of several identical lines with tunable intensities and wavelengths) was investigated by the excitation of multifrequency gratings in the acoustooptical deflector that was placed in the dispersive resonator [1,13,14]. However, it is impossible to approximate continuous spectra with a series of discrete spectral lines separated by narrow spaces in above multifrequency lasers because the destructive influence of both additional spectral components[13] and of intercombinative interactions[15] increases essentially.

To do it, we proposed and sequentially consider (theoretically and experimentally) the methods based on quick shaping the dispersive resonator spectral function (Sp.-Func.) in accordance with the necessary spectrum shape. We shall use the other possibilities of the acoustooptic devices that allow to flexibly shape in a controlled way the complex spatial distribution of the transmittance inside the laser cavity.

This investigation is also important from the viewpoint of general studies of dispersive resonator properties and elucidation of regularities of the spectrum evolution during the laser generation.

Until this moment only the direct problems for the dispersive cavities were studied, including a set of our works. In particular, we have sequentially studied methods for evaluation of their spectrum width [16,17], the influence on it of different factors (such as the resonator configuration [16,18], mirror spatial profile [19], dynamic lens [20], and random irregularities in active media [21]), features of selection in ring dispersive cavities (including ones with dispersion devices combining properties of the input-output gate with different reflectivity [22]), the interference effects in the special case of strongly coupled dispersive resonators [23], explanation of unusual experimental effects in lasers with dispersive cavities without engaging aberration effects [24] etc. Yet no attention was attracted to inverse problems.

In addition, there are important unsolved problems in understanding of spectrum evolution, in particular, it is not clear how the active medium influences the generation spectrum and its evolution. It was shown (see, for example, [25] etc.), that under the condition of homogeneous broadening of the gain line (only such media are effective in tunable lasers) the spectrum is narrowed proportionally to a square root from the number of the light passes through the resonator and this regularity does not depend on the filter arrangement. At the same time in lasers with the selective devices possessing the angular dispersion (gratings and prisms), in comparison with others, there is the key distinction arising from specificity of the selection mechanism: the change of an equivalent declination during change of a wavelength inside limits of the oscillation spectrum is conducted with the displacement of the field in the transverse direction and, hence, with field displacement in the active medium.

In Section 2 we shall consider the basic equations necessary for studying the processes in proposed lasers. In section 3 we shall deal with the formation of Sp.-Func. of two slightly diffractively coupled dispersive resonators, and clarify the main properties of Sp.-Func. formation. In Section 4 we develop the algorithm for the Sp.-Func. formation (in case of continuous spectrum synthesis) when its moments are given. In Section 5 we shall consider the new regularities of the spectrum evolution which are connected with influence of "the spatial heterogeneous broadening" in the active medium. In Section 6 the algorithm of spectra synthesis is modified to include the processes occurring in active medium. In Sec.7 results of experimental study of the laser equipped with a novel control element are shown.

2 Basic Equations Describing Spectrum Formation and Its Evolution

The time-dependent behavior of laser spectral characteristics is mainly determined by two processes: 1) the competition of different components of the spectrum in the same active medium (with homogeneous broadening); 2) the wavelength-dependent selection of different components in the dispersive resonator, which is characterized by its Sp.-Func. (i.e. the dependence of quality factor on the wavelength). Therefore, the complex spectrum formation may be studied only by the analysis of the basic problem of laser generation. Below we shall consider the contribution of each of these two processes.

At the same time, it is necessary to note that in the short-pulse lasers (with duration ~10 ns), where experimentally observable quantities are averaged per one pulse, the competition processes are not very significant and cannot substantially change the spectrum shape (the averaged quantities are approximately determined by half from the total number η of light passes through the resonator that is η_s~1,5-3,5 for η~3-7). Therefore to control the spectrum shape we chose to control the Sp.-Func. of the dispersive resonator. We shall elaborate the algorithm of tuning the Sp.-Func. (i.e. finding the values of control parameters implementing the necessary shape of Sp.-Func.) with its subsequent improvement to include the competition processes in the active medium (including nonlinear ones).

Generally, it is possible to change the Sp.-Func. shape by separate control of the amplitude norm and the phase of the spatial distribution (below N-Dist. and P-Dist.) of the light transmission factor in the cross-section of dispersive resonator. To do this in the tunable laser (Fig.1b, 8) we introduce a novel control element 2, which includes an acoustooptical spatial modulator 3 (to change N-Dist.) and a lens telescope 4 (to tune P-Dist. by simply changing the distance between lenses). The specific details of the experiment setup will be discussed in Section 7.

Without loss of generality we assume that electromagnetic field depends only on two spatial coordinates in the plane of the dispersion of grating 1. Such a model will be sufficient for an adequate study of the wavelength selection properties in this system.

Deriving the equation set for the laser which resonator contains devices with the angular dispersion, we shall take into account that here the frequency selection is defined by the increase of losses in the course of the gradual displacement of the field to the mirror edge. Thus spatial distributions of modes (already even longitudinal), having different wavelengths, will be different in the plane of dispersion of such resonators. The consequence of this can be unequal "saturation" of the active medium by them. In some sense it is equivalent to the occurrence of the partial heterogeneous broadening which influence can be expected at nonlinear stages of generation.

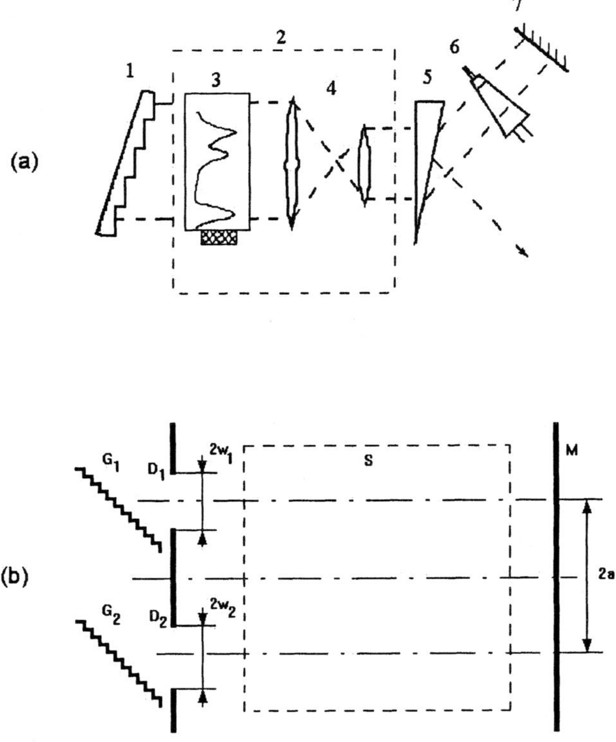

Figure 1: (a) Dispersive resonator with the tunable shape of the spectral function. (Here are grating 1, control element 2, acoustooptical spatial modulator 3, lens telescope 4, prism expander 5, cell with the dye solution 6, mirror 7), (b) The model of two diffractively coupled dispersive resonator.

The simple model, which would allow to modify the equations for such systems and to take into account such above processes, is following.

Let's consider the laser (Fig. 1) which active medium 6 is located near the mirror 7 and has the length significantly smaller than the cavity length. Such geometry, typical for many actual systems (for example, for lasers on dyes and F centers), will enable us to significantly simplify the problem. Without sacrificing the generality of our study, we assume that the active medium is located in the plane of the mirror, the laser transition is homogeneously broadened and is defined by the four-level scheme typical for the majority of tunable lasers (on electron-phonon transitions, on dyes[26] etc.). Inside the dispersive cavity there are a spatial modulator and other focusing units and diaphragms. The modes of the dispersive cavity $u_{nj}(x)$ are eigenfunctions of the following integral equation[16]

$$\Lambda u(x) = \int_{-\infty}^{\infty} K(x,x_1)u(x_1)dx_1 \,, \tag{1}$$

where $K(x,x_1)$ is the kernel describing the field transformation in linear units (its explicit form will be given below at realization of concrete calculations), and $\Lambda = e^{-\gamma}$ is the eigenvalue describing losses of modes and extra phase shift.

The field $E(x,t)$ before the entrance to the active medium in the direction to the mirror 6 is expanded in modes $u_{nj}(x)$ of the dispersive cavity. Then $E(x,t)$ is

$$E(x,t) = \sum_{j} e^{i\omega_j t} \sum_{n} E_{nj}(t)\, u_{nj}(x), \tag{2}$$

where ω_j are the frequencies of longitudinal modes, and $E_{nj}(t)$ are the expansion coefficients with slow time variation which take into account also the distinction of frequencies of longitudinal modes and modes of higher order. In order to obtain the set of differential equations for the electromagnetic field we shall take advantage of simple procedure of deriving them from equations written in terms of finite differences. After one pass through the resonator (during the time τ_l) the field is defined as follows

$$E(x,t+\tau_l) = \sum_{j} e^{iw_j(t+\tau_l)} \sum_{n} E_{nj}(t) \int_{-\infty}^{\infty} K(x,x_1)u_{nj}(x_1)e^{2\sigma N_2(x_1,t)\Delta z}\, dx_1, \tag{3}$$

where Δz is the length of the active medium, $N_2(x,t)$ is the population difference of the laser transition, σ - cross-section of target of laser active centers. The integral equation for cavity modes is not Hermitian and can be even asymmetric. Therefore for expansion of a field in (3) we shall use the sets of eigenfunctions $\{u_{nj}(x)\}$ and the ones conjugated to them $\{v_{nj}(x)\}$ (that are the solution of the equation conjugated to (1)) which have the same eigenvalues and are normalized in the following sense

$$\int u_{nj}(x) \, v_{kj}(x) \, dx = \delta_{nk}$$

After not very complex transformations, we obtain the expression for the change of coefficients in (2) after one pass

$$E_{kj}(t + \tau_l) = \sum_n E_{nj}(t) \int u_{nj}(x_1) \, e^{2\sigma N_2(x_1,t)\Delta z - \gamma_{nj}} v_{kj}(x_1) dx_1 , \tag{4}$$

Constructing the finite difference $(E_{kj}(t+\tau_l)-E_{kj}(t))/\tau_l$ and using traditional approaches (going from the finite difference to the derivative and assuming that coefficients in (4) do not change greatly during one radiation pass through the resonator) we obtain the equation set describing evolution of the field variables

$$\frac{dE_{kj}(t)}{dt} = \frac{1}{\tau_l}\left[\sum_n E_{nj}(t)2\sigma\Delta z \int N_2(x,t) \, u_{nj}(x)v_{kj}(x)dx - \gamma_{kj}E_{kj}(t) \right] \tag{5}$$

For completeness the set (5) needs to be supplemented by material equations and by function of dependence of γ_{kj} on the wavelength. The material equations for $N_2(x,t)$ can be obtained by known methods: rigorously, using a density matrix of quantum mechanics, or phenomenologically, using the temporal equations for populations or their equivalent representation for the amplification constant of the active lamina. The only complication consists in necessity of taking into account the dependence of the amplification constant on the cross-space variables. The concrete relations will be presented below in sections 5,6.

Parameter γ_{kj} in (5) defines the field attenuation in the resonator and includes common constant losses and their diffraction part, which depends on the wavelength and is defined by the dispersive cavity's Sp.-Func. To obtain Sp.-Func. of a resonator with the given configuration (for example, Fig.1a) it is necessary to find the dependence of eigenvalues $\Lambda(k)$ on the wavevector k in the integral equation (1).

For the convenience of analysis we rewrite Eq.(1) by transferring the plane of reference in a plane of diaphragms (modulator) (Fig. 1a) and by detailing its kernel: [16]

$$\Lambda(k)u(\xi) = \int_{-\infty}^{+\infty} K(\xi,\xi_1)N(\xi_1)P(\xi_1)W(\xi_1,k)u(\xi_1)d\xi_1 \tag{6}$$

where $u(\xi)$ is the electromagnetic field in the diaphragm plane for the wave incident on the grating (gratings); $W(\xi_1, k)$ is the function describing the wave inclination by the grating that depends on k (λ); N and P are N-Dist. and P-Dist. modeling the action of control element 2; $K(\xi,\xi_1)$ is the function describing the propagation of the wave field through of the rest of the resonator; $\xi=x(k/L_e)^{1/2}$ is the dimensionless coordinate measured from the optical axis; L_e is the diffractive length of the resonator.

$|A(\Delta k)|^2$ permits to estimate the selective properties of the system. However, even the direct task (i.e. the evaluation of $|A(\Delta k)|^2$ by repetitive solving Eq.(6) for different values k) is very difficult for solving in arbitrary case (for complex functions N and P). Therefore the inverse task in the full statement (i.e. the determination of unknown functions N and P from Eq.(6) when the function $G(k)=|A_0(\Delta k)|^2$ is known) can be solved only by the complex algorithms of the iterative refinement of the initial tunable parameters of El.2 with the use of the verification of the convergence of $|A_0(\Delta k)|^2$ to $G(k)$ at each step in case when the direct task may be solved.

To facilitate the analysis we first consider the case that still allows to study the direct task in the full statement (section 3), and then we shall discover the approximate solutions in more general case (section 4).

3   Spectral Function of Dispersive Resonators with Diffraction Coupling

Let the resonator has N-Dist. in the form of two identical diaphragms with rectangular profile. This problem is equivalent to that of two diffractively coupled dispersive resonators (Fig.1b). Assume that resonators, each include a grating G_i and a diaphragm D_i, have the common optical system S and the infinite mirror M. The distance between the optical axis of S and the center of the diaphragm D_i is equal to d. The coupling between the resonators occurs during the electromagnetic wave propagation because of diffractive "streaming" of the field from one resonator to another. In this case the function $K(\xi,\xi_i)$ in Eq.(6) is:

$$K(\xi,\xi_1) = \frac{\Gamma}{\sqrt{2\pi\ i}}\exp(-i\xi\xi_1 + ig(\xi^2 + \xi_1^2)/2) \qquad (7)$$

where Γ is transmittance of the optical system S, g is its equivalent configuration parameter defined per two passes through the system. For the simple resonators in study with the uniform diaphragms we represent the influence of the functions N, P, and W by single that $N(\xi, k)$ describing the diaphragm transmittance profile and the grating action:

$$N(\xi,k) = t_j \exp(-i\gamma_j\xi), \ \xi \in [\alpha - \beta_j, \alpha + \beta_j], \ j = 1,2 \qquad (8)$$

where $\beta_i = w_i(k/L_e)^{1/2}$ is the equivalent dimensionless aperture parameter of the diaphragm D_i with halfwidth w_i; $\alpha = d(k/L_e)^{1/2}$ is that for the distance between the diaphragms, which is equal to $2d$; $\gamma_i = \pi\Delta k(L_e/k^3)^{1/2}D_{\phi i}$ is the dimensionless parameter for the wave inclination by the grating (here Δk is the detuning of the wavevector from its value for what the grating does not introduce the additional wave front inclination in the resonator; $D_{\phi i}$ is the angular dispersion of the grating G_j; $\gamma_1 = \gamma_2$ if the grating is common); t_1, t_2 are the terms that account for possible differences in the transmission factors of the diaphragms.

Results obtained in [16,19] allow to assume Sp.-Func. to depend on the diaphragm transmittance profile in the region of $\beta > 1$. Therefore, below we have concentrated on the numerical study of Sp.-Func. in the regions both of high apertures, where the coupling of resonators is weak, and of their intermediate values, as opposed to [23], where the interference effects have been analytically studied for $\beta < 1$ and high coupling.

To solve Eq.(6) it has been approximated by the matrix equation. Its eigenvalues have been determined by the LR algorithm with the standard subroutine COMLR from the Sandia Library. The most essential results of numerical calculations are shown in Fig.2. We studied the dependence of Sp.-Func. on the aperture parameter β (the increase of which in case of the flat-mirror resonator synchronously leads to the decrease of the diffractive coupling between resonators) and on the resonator configuration given by the value of Re g (Re $g=1$ corresponds to the flat-mirror configuration).

Fig.2a shows the Sp.-Func. shape evolution as the value of β is changed for Re $g \neq 1$ (since Sp.-Func. is symmetrical we show it only for $\Delta k > 0$). In the case of sufficiently high coupling ($\beta = 0.5$ in Fig.2a) there is only one central peak at $\Delta k = 0$ independently of system configurations (Re g), that is due to the predominate influence of the interference effects as in [23]. In going to the region $\beta > 1$ N-Dist. starts to influence the Sp.-Func. shape. As β increases, initially two separated maxima appear alongside the central peak. Then the central maximum amplitude quickly decreases, and the positions of the lateral ones cease to vary (become stable).

Fig.2b shows the influence of Re g for some fixed $\beta > 1$. For Re $g=1$ Sp.-Func. has one maximum at $\Delta k = 0$. When the configuration departs from that there exist two symmetrical maxima which positions depend on $|g - 1|$.

However, the particular value of β for resonators to be weakly coupled may depend on g. When Re g essentially departs from 1 and falls into the region of Re $g \cong 0$ (Fig.2c) or of unstable resonators, the central maximum can appear again that implies the increase of the resonator coupling.

Possibilities to calculate Sp.-Func. in the full statement of the problem is practically exhausted by such simple profiles of diaphragms like one discussed above. However, the obtained results permit to draw conclusions about the properties of more complex resonators and to formulate (in Sec.4) the simple model for the approximate evaluation of resonator parameters from the given Sp.-Func.

Let us find now the analytical relation for the dependence of the Sp.-Func. maximum position (Fig.2b) both on P-Dist. (Re g) and on the diaphragm characteristics. To do this we assume the profile of the intracavity diaphragm to be Gaussian. This assumption is not restrictive since the numerical analysis of Eq.(6) for the coupled resonators with different profiles of the mirror reflectance (including parabolic one) has shown, that the influence of the exact profile of partial diaphragms becomes insignificant with the increase of the value of $|g-1|$.

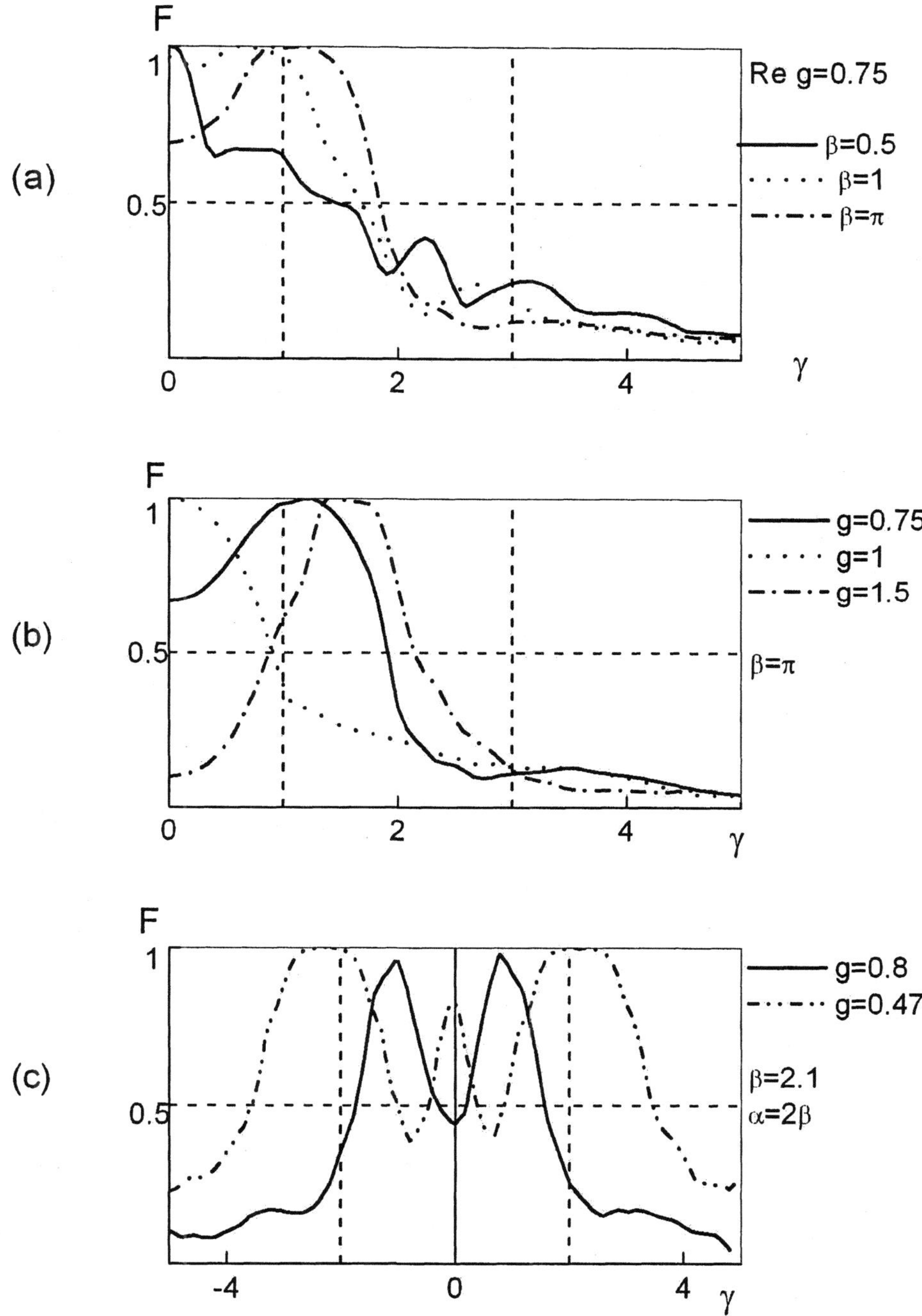

Figure 2: Calculated spectral function of the coupled dispersive resonators: (a) dependence of its shape on the aperture parameter β for fixed resonator configuration; (b) dependence of its shape on the resonator configuration Re g for fixed β; (c) dependence of its shape on the resonator configuration Re g that illustrate the increase of the resonators coupling.

Using the Gaussian beams optics, [16] we obtain the following expression for Sp.-Func. of the resonator with the single diaphragm displaced from the axis of the optical system at the distance d:

$$F(\Delta k) = |\Lambda(0)|^2 \exp(\text{Im}\frac{s^2}{g_1 - 1}) \qquad (9)$$

where $s = \gamma/2 + \alpha(g-1)$; $g_1 = g + i\beta^{-2}$, $g = 2AD-1$ (A, B, C and D are terms of the transfer matrix of the optical system S).

As it follows from Ex.(9), with the departure of Re g from 1 the asymmetry of the relative positions of the axes of the optical system and the diaphragm gives rise to the shift of the position of Sp.-Func. peak. The shift can be written as:

$$\delta = \left(\frac{\Delta k}{k}\right)_M = \frac{2d}{\pi D_\phi}\frac{\text{Im}((g-1)(g_1^*-1))}{\text{Im}(L_e(g_1^*-1))} \qquad (10)$$

The substitution of the relation $L_e = 2BD$ and the parameters of the diaphragm and optical system with the real transfer matrix gives the simple expression:

$$\delta = d\frac{2C}{\pi D_\varphi D} \qquad (11)$$

Conclusions of Ex.(11) are that the detuning δ is independent of the diaphragm parameters and its value is determined by the diaphragm displacement from the optical axis. The peak positions of spectral functions in Fig.2 are in very close agreement with that calculated from Ex.(11). More general expressions for the wavelength shift in the case of the optical system displacement in the dispersive cavity are investigated in [24]. They allow to find necessary parameters for the modulator in the arbitrary case.

4   Synthethis of Dispersive Resonator with Spectral Function Given by Its Moments

Therefore in the intermediate region one can create the conditions when the laser spectrum is determined by the intracavity diaphragm profile, and the change inside the individual section of the diaphragm acts locally upon the spectrum shape. Under these conditions formation of the spectrum shape can be fulfilled in two stages. The first, giving the spectrum scale (the difference in tuning wavelengths of different sections), is to form P-Dist. and the second, giving the spectrum shape, is to form N-Dist. As we show below, these two stages are not independent and therefore the algorithm of the spectrum formation is essentially nonlinear.

Implementing the system one can use either the continuous N-Dist. or the discrete one. In this section we focused on the latter case for real calculations to be possible, i.e. we use N-Dist. in the form of 2m+1 sections, treating each as an individual partial resonator. Then we assume each section to have smoothed-out borders (for example, with parabolic distribution) and approximate Sp.-Func. of partial resonators with Ex.(9). In this case Sp.-Func. can be approximately written as:

$$F(k) = \sum_i a_i F_i(k - \Delta k_i) \qquad (12)$$

where each F_i is defined by the difference Δk_i that describes the wavevector tuning and by the half-width σ:

$$\delta_i = \frac{\Delta k_i}{k_0} = \frac{k}{\pi D_\phi} \frac{x\Delta}{f_2^2} \qquad (13a)$$

$$\sigma = \frac{k}{\pi D_\phi} \left(\frac{1}{k^2 w^2} + \left(\frac{w\Delta}{f_2^2} \right)^2 \right)^{1/2} \qquad (13b)$$

Here f_2 is the greatest focal length of the telescope lenses and Δ is the value of the lens telescope detuning (so the distance between the lenses is equal to $f_1 + f_2 + \Delta$).

Equation (12) can be used for the determination of the system parameters if the function $F(k)$ is given. Unknown terms are the controlled parameters of resonator, that are the amplitudes a_i of the individual section's transmittance factor and the detuning Δ of the lens telescope. If it is necessary, the distance x between lateral diaphragms and their half-widths w can also serve as unknown terms. However, one should remember that the interval of their change is limited from below by the requirement of the weak diffractive coupling between partial resonators, and from above by the aperture of El.2 and by the minimum number of diaphragms that still permits to form the spectrum with the necessary complexity.

Let us dwell on the role of the variable Δ influencing both the distances between the partial resonators Sp.-Func.'s and their half-width. From this viewpoint the following relation is remarkable:

$$y = 2\left(\frac{\delta}{\sigma}\right)^2 = 2\left(\frac{x\Delta}{f_2}\right)^2 \bigg/ \left(\frac{1}{k^2 w^2} + \left(\frac{\Delta w}{f_2^2}\right)^2\right) \qquad (14)$$

which defines the overlap of Sp.-Func.'s of neighboring diaphragms. For the large values of Δ the value of y is approximately equal to $2(x/w)$, and the functions F_i are separated. In this region the variation of Δ, when the other parameters are fixed, induces a similar change of the spectrum. When both terms in Eq.(13b) are comparable, y depends on Δ and the variation of Δ leads to the dissimilar change of the spectrum shape. In the case of formation of continuous spectra we are forced to work in the last region, that complicates the synthesis algorithm, which becomes nonlinear.

Since real systems have finite number of controlled parameters, it makes sense to separate several characteristics that describe the main features of a continuous spectrum. Here we chose several lowest moments of the spectral distribution [27], namely, the first absolute moment:

$$m_1 = \int (\lambda - \lambda_t)I(\lambda)\mathrm{d}\lambda$$

which determines the change of the mean wavelength relative to the tuning wavelength λ_t because of asymmetry of the spectrum shape, and several central moments:

$$M_k = \int (\lambda - m_1)^k I(\lambda)\mathrm{d}\lambda$$

determining, for example, the width, the asymmetry, and the kurtosis of spectrum. In this case to synthesize the laser spectrum in accordance with the given shape is to find the resonator parameters that provide the closest matching for the given values of moments.

The algorithm of the solution becomes more simple by introduction in the modulator of extra degrees of freedom, which ensure the maintenance of zero value of the first moment during the spectrum tuning ($m_1 = 0$). This condition becomes possible generally for $2n+1 \geq 5$, and is automatically fulfilled for $2n+1 \geq 3$ in the case of symmetric spectra. Instead of the analysis of the set of the nonlinear equations, the problem is reduced to the solution of one nonlinear equation relative to Δ with the subsequent determination of amplitude distribution from the set of linear equations. We shall give explicit expressions for this equation set with the concise presentation of its deriving.

Going to the dimensionless moments $m_k = M_k/(M_2)^{k/2}$ (in order to separate the parameters of Sp.-Func. that are independent of its width) we derive the following equation set (without loss of generality one can assume F to be normalized):

$$m_{2r} = \frac{1}{z^r} \sum_{j=0}^{r} T_{2r}^{2j} y^j b_{2j} \qquad (15a)$$

$$m_{2r+1} = \frac{1}{z^{(2r+1)/2}} \sum_{j=0}^{r} T_{2r+1}^{2j+1} y^{(2j+1)/2} b_{2j+1} \qquad (15b)$$

$$T_{2r}^{2j} = (2(r-j)-1)!! C_{2r}^{2j}; T_{2r+1}^{2j+1} = (2(r-j)-1)!! C_{2r+1}^{2j+1} \qquad (15c)$$

where $z = 1 + y\, b_2$, and

$$b_{2k} = \sum_{i=1}^{m} i^{2k}(a_{-i} + a_i), b_{2k+1} = \sum_{i=1}^{m} i^{2k+1}(a_{-i} - a_i) \qquad (15d)$$

Thus even (odd) moments will be bound only to even (odd) part of N-Dist.

The equation set (15) is infinite. We restrict it accordingly to the number of moments and resonator parameters that is necessary for the chosen approximation. This set, unlike the starting one, is linear in $\{b_i\}$. Let us reduce it by rearrangement to one nonlinear equation in Δ. To do this first we solve relations for even moments in linear variables:

$$1/(2k-1)!!\, b_{2k} y^k = \sum_{i=0}^{k} H_i^k \mu_{2i} z^i \qquad (16a)$$

where $\mu_{2i} = m_{2i}/(2i-1)!!$. Substituting relations (15a) in (16) and equating expressions by identical degrees y^k in both parts, we obtain the equation set for unknown coefficients H

$$\sum_{i=m}^{k} \frac{(2m-1)!!}{(2i-1)!!} T_{2i}^{2m} H_i^k = 0, \text{ if } m \leq k-1 \qquad (16b)$$

$$H_k^k = 1 \qquad (16c)$$

which has the simple triangular shape. Solving sequentially equations in the direction of the decrease of m and using the method of mathematical induction, we discover

$$H_i^k = (-1)^{k-i} C_k^i \qquad (16d)$$

Then, by taking into account that only n variables b_i are independent because of the finite number of diaphragms, i.e.

$$b_{2n+2} = \sum_{k=1}^{n} D_k b_{2k}$$

[D_k can be determined from Eq.(15d), for example, if $2n+1=5$ they are $D_1=-4$, $D_2=5$] one can obtain the following equation set

$$\sum_{i=0}^{n+1} z^i \mu_{2i} \sum_{j=\max(1,i)}^{n+1} \frac{(2j-1)!!}{(2(m+1)-1)!!} H_i^j D_j y^{n+1-j} = 0 \tag{17a}$$

$$M_2 = \sigma^2(\Delta) z(\Delta) \tag{17b}$$

for $z(\Delta)$, $y(\Delta)$, where $D_{n+1}=-1$. The first equation (17a) defines z as the implicit function of y independently of the spectrum width. The second one (17b) gives the specific spectrum width. One can obtain one nonlinear equation in Δ (which explicit form is not presented by us) eliminating $z(\Delta)$ from (17a) and taking into account (13, 14).

Solving Eq.(17) together with linear Eq.(15a), (14) one can find the resonator parameters Δ and $\{a_i\}$. Because of the different character of the influence of Δ on spectrum, the procedure of the determination of laser resonator parameters is different in regions of similarity and dissimilarity. In the similarity region the equation set is separated with respect to $\{a_i\}$ and Δ. The spectrum shape is defined only by N-Dist. ($\{a_i\}$) and the ratio x/w, and the necessary spectrum width is independently determined by the choice of Δ.

In the dissimilarity region the task is more complex. As the analysis of this case has shown, for solving the equation in Δ (or the equivalent equation set (13, 14, 17)) it is efficient to use the iterative procedure with the following correction relation (refinement) at every step

$$\Delta_{r+1} = \Delta_r R, \quad R = \frac{M_2}{M_2^r}$$

until $R \cong 1$ is achieved (this corresponds to the tuning of obtained solution to the given spectrum width). The iteration number needed to obtain the solution is equal to $3\div7$.

Examples of synthesized spectral function of the dispersive resonator are shown in Fig.3. For the same spectrum width the quality of approximation depends on the ratio x/w. When the required spectrum width is increased, the overlap factor of Sp.-Func. decreases and the quality of approximation deteriorates (Fig.4).

From the analysis of solutions for the synthesis of the spectra that have the same shape but essentially different widths, both N-Dist. and the functional species of the dependence of Δ on the spectrum width are different because of the nonlinear influence of the variable Δ on the partial functions widths. We demonstrate this by a specific example of Sp.-Func. with $\mu_3=-0.194$, $\mu_4=0.722$, $\mu_6=0.447$ ($w=0.77$mm, $x=2w$, $2n+1=5$) for two spectra which squares of the widths differ in ten times ($M_2=45$ and 450). We found the following N-Dist. in the first case:

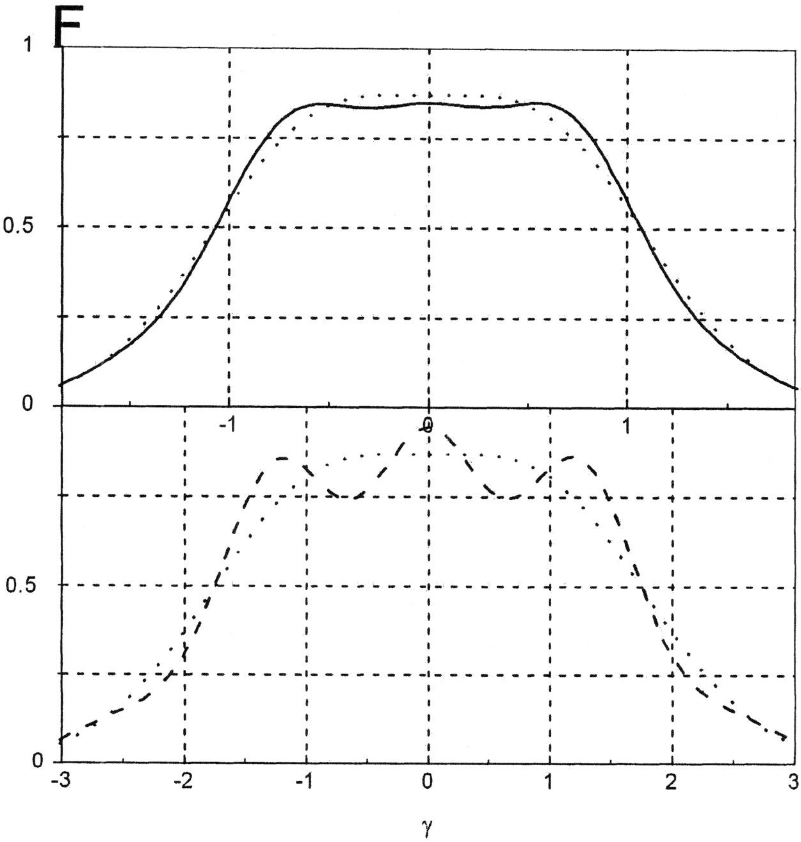

Figure 3: Spectral function of the dispersive resonator calculated with the set of equations (15, 16, and 17) from the given moments. Dotted line - needed spectral function, solid - calculated with the ratio $x/w=2$, dashed - $x/w=4$.

$a_{-2}=0.22$, $a_{-1}=0.89$, $a_0=0.53$, $a_1=1.18$, $a_2=0.08$

and the following for the second case:

$a_{-2}=0.21$, $a_{-1}=1.00$, $a_0=0.77$, $a_1=1.27$, $a_2=0.08$.

They are essentially different (particularly a_{-1}, a_0, a_1).

The analysis showed that the proposed method of the resonator parameters calculation from the given moments permitted the construction of spectra with essentially different width (it is possible to change it up to $20 \div 30$ times) when the values of x and w are constant.

The structure of equation set (15, 16, 17) and the algorithm of its solution are general enough and do not depend on a specific choice of symmetric Sp.-Func.'s of partial resonators in (12). The Gaussian functions were chosen for realization of concrete calculations in the present section. In the case of going to the set of symmetric functions, which correspond to other shape of sound pulses in the spatial modulator (rectangular ones etc.), it is necessary only to change expressions for constant coefficients T in (15c) and H in (16d). However, their values, and Sp.-Func.'s for partial resonators are necessary to find by the numerical methods.

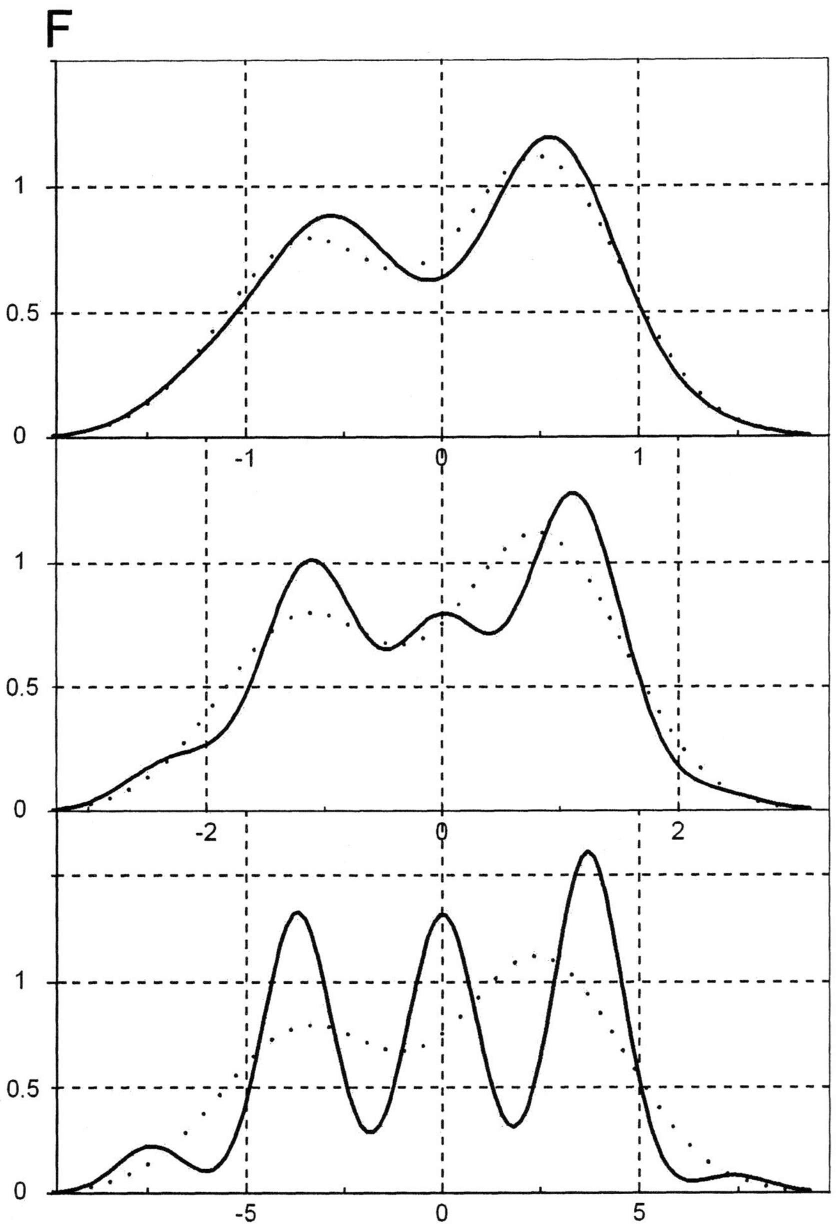

Figure 4: Influence of the partial diaphragm parameters on the spectral function of the dispersive resonator calculated with the set of equations (15, 16, and 17) from the given moments in the non-similarity region. Dashed line - needed function, solid - calculated.

5 Single Line Spectrum Evolution with Distinctions of Modes' Spatial Distributions Taken into Account.

In sections 5,6 we shall consider how the laser spectrum is affected by variation of spatial distributions of mode's fields inside limits of the generation spectral line. This mechanism is specific for resonators with the angular dispersion. Below two practically important cases are studied. They will allow us to formulate corrections to algorithm of the spectrum shape tuning.

In the cavities with units having the angular dispersion there are two mechanisms of selection losses [19]. First of them, which takes place for the rectangular shape of the resonator transmittance, is the diffraction mechanism. In this case the mode field maximum slips on the mirror edge during the wavelength change from the spectral line center to its border. It gives rise in loss increase because of radiating a part of the mode field outside the mirror. At the same time there is gradual deformation of the mode distribution.

The second mechanism is implemented for mirrors with the parabolic profile of reflectivity or with Gaussian that [19]. The field maximum gradually slips on the border similar to the previous case. However the field distribution varies much less. Here the selection (increasing losses for border wavelengths) is defined by the field maximum shift in regions with the smaller reflection. This case is implemented in many actual systems. The zone of excitation practically has the Gaussian profile in lasers with the longitudinal laser pumping. Besides, the resonators, which mirrors have the rounded parabolic profile, are frequently used for improvement of the mode composition (for selection of dominant modes), and as alternative to unstable resonators (see, for example, [28,29,30]).

We shall consider the second case, since it is practically important, and allows essentially to simplify the calculation procedure due to the use of the model of mirrors with Gaussian diaphragms, for which there are analytical expressions for modes. The essential feature of such model is the regular relation of losses of modes of the different order (when the mode of the lowest order always has the lowest losses). Therefore in the region of the aperture parameter $\beta=(\pi w^2/(\lambda l))^{1/2} <10$ (with sufficient differences in losses of modes) and with sufficient duration of the oscillation linear stage there will be only longitudinal modes.

The resonator will be considered to contain an autocollimation diffraction grating and optical devices of Gaussian optics ensuring the single spectrum line oscillation. The kernel of the integrated equation (1) for such resonator is symmetric if the reference plane is on the output mirror [16,24]:

$$K(\xi_1,\xi)=(-i/(2\pi))^{1/2} \exp\{i[f(\xi_1+\xi)+(g\xi_1^2+g\xi^2-2\xi_1\xi)/2]\},$$

where the following symbols are used:

$$g=A_s, \quad f=(2\pi B_s/\lambda)^{1/2}D_a\lambda_o(\Delta k/k_o)2, \quad \xi=x(k/B_s)^{1/2}; \quad A_s=1+2CB; \quad B_s=2BD;$$

A, B, C, D are elements of the matrix of the resonator's optical system containing the Gaussian diaphragm; D_a is an angular dispersion of the grating; $\xi=x(k/B_s)^{1/2}$ is a dimensionless space variable.

The field distributions of longitudinal modes of such resonator and their eigenvalues look like

$$u_0 = \sqrt{\frac{2\sqrt{g^2-1}}{i\pi}}\ \exp\left[i\sqrt{g^2-1}(\xi+\frac{f}{g-1})^2/2\right];$$

$$\Lambda_0 = M^{1/2}\exp\left[-i\frac{f^2}{g-1}\right]; \quad M = g+\sqrt{g^2-1},$$

where the values of "tunings" Δk are chosen from requirement of mode field reproducibility, and signs before radicals are determined from requirement of field decrease when ξ tends to infinity. The complex spatial displacement of mode fields (their amplitudes and phases) is characterized by the expression $f/(g-1)$.

Using such approach and the relation $q_j(t)\sim |E|^2$, we substitute photon numbers for field variables in the set of equations (5). In this case the equations acquire the form which takes into account the dependence on the transversal coordinate x of spatial distributions of both energy of modes and the population $N_2(x,t)$ of the excited state of four-level working transition

$$\frac{dq_j(t)}{dt} = 2q_j(t)\left[\frac{\sigma\Delta z}{\tau_l}\int N_2(x,t)Z_j^q(x)dx - \frac{\gamma+\mathrm{Re}\,\gamma_{0j}}{\tau_l}\right], \tag{18а}$$

$$\frac{dN_2(x,t)}{dt} = W_p(t)(N_T-N_2(x,t))-U\cdot N_2(x,t)\sum_j Z_j^N(x)q_j(t) - \frac{N_2(x,t)}{\tau} \tag{18б}$$

where q_j is the photon number of the mode j, U is the Einstein coefficient for induced transitions ($U=\sigma c/V$, where c is the light velocity, V is the mode volume in the active medium), τ is the time of spontaneous emission, l is the resonator length, $\gamma = -\ln R$ is the logarithmic losses per one pass through the resonator that is the same for all modes, $\mathrm{Re}\,\gamma_{0j}$ is the selective losses of the mode j, $W_p(t)$ is the pumping function, N_T is total number of laser centers in the active medium. It is necessary to emphasize that in nonlinear terms the "connection" coefficients in the equations for photon numbers and for the inverted population are different (the role of this factor has been studied specially in [33]).

$$Z_j^N(x) = u_{0j}(x)u_{0j}(x)^*; Z_j^q(x) = 2\operatorname{Re}[u_{0j}(x)u_{0j}(x)]$$

The set (18) was analyzed numerically with the use of the Euler generalized method for the temporal derivative approximation and with parameters that are typical for the dye lasers (for example, for solution of Rhodamin 6G in ethanol) pumped by the short-pulse laser. The number of modes was selected from an interval 20 - 60. The most typical effects are given in Fig. 5,6,7.

The values of the aperture parameter β and the excess of the pumping energy over the threshold one $\rho=W/W_0$ were varied. The main conclusion is that taking into consideration the spatial structure variation leads to the essential change of spectrum kinetics during the laser pulse for $\beta \geq 3,5$ and $\rho > \rho_0$ (ρ_0 depends on β and diminishes with its magnification). Fig. 5 illustrates typical features of this for $\beta=4$.

For $\rho=5$ (Fig. 5 a) the spectrum evolves in the usual way having the same shape and being narrowed down proportionally to the square root from the number of radiation passes through the resonator. For $\rho=15$ (Fig. 5 c) narrowing the spectrum occurs during the first three passes as well as in the previous case. During the fourth pass there is a rise in wings, to the sixth pass the spectrum is converted in double-peak one, and during the seventh pass it is transformed in the rectangular one. To the tenth pass the spectrum is again converted in Gaussian with gradual narrowing. These phases make the first cycle of dynamics. The curious case is demonstrated by Fig. 5 b for $\rho=10$. The spectrum is in steady-state after the second pass and keeps the constant shape and width during the pulse. (Here and below we use dimensionless variables for the wavelength and the spectrum in relation to the width σ_0 of the dispersive cavity Sp.Func.)

The experimentally observable parameters are characterized by the mean spectrum and the mean width of the spectrum calculated per first cycle. They are given in Fig. 6,7. Fig. 6a illustrates the dependence of the mean spectrum on β for $\rho=15$. Despite similarity of the first cycle phases, the mean spectrum changes its shape as it depends also on the kinetics of the total intensity in the course of the pulse, and the kinetics in turn depends on β. In Fig. 6b the mean spectrum for $\beta=4$ is shown depending on the excess ρ of the pumping energy over the threshold one which supplies effects represented in Fig.5.

In Fig. 7a the temporal kinetics of the spectrum width is shown for different aperture parameters β in the case of $\rho=15$. Here the dotted line gives the dependence according to results [25] that allows obviously to get regularities obtained by us. It is necessary to note that the nonlinear processes develop fast enough, which is emphasized by occurrence of distinctive springs. In Fig. 7b the dependencies on β of the mean width of the spectrum are given. For $\rho=15$ in the interval of $\beta=3 \div 4$ there is a sharp enough deceleration of spectrum narrowing caused by nonlinear processes in the active medium. And though the nonlinear dynamics (Fig. 5в) develops beginning with $\beta\sim4$, the deceleration of the rate of spectrum narrowing occurs essentially earlier.

The obtained effects are similar to effects taking place in solid lasers with the spectroscopic–heterogeneous broadening of working transition. In our case the working transition remains homogeneously broadened. The role of heterogeneous broadening plays the variations of the mode spatial structure inside limits of the oscillation spectrum, and the role of the width of the homogeneously broadened spectral line plays the mode spatial distribution width. The physical explanation of the spectral dynamics in Fig. 5c is

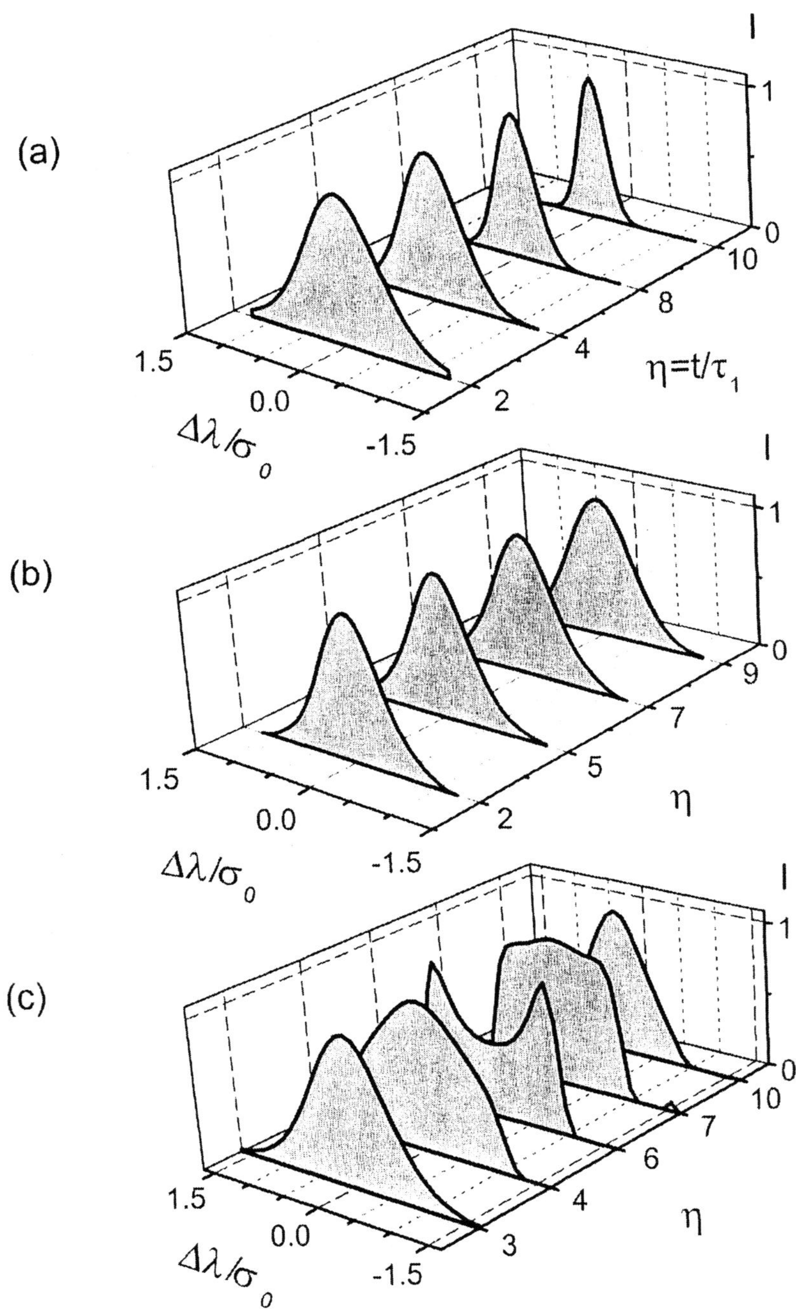

Figure 5: Dependence of tunable laser spectral dynamics (calculated with the set of equations (18)) on the time ($\eta=t/\tau_l$) for the aperture parameter $\beta=4$ and different pumping energy W: (a) $\rho=W/W_0=5$ (W_0 is the threshold pumping energy), (b) $\rho=10$, and (c) $\rho=15$.

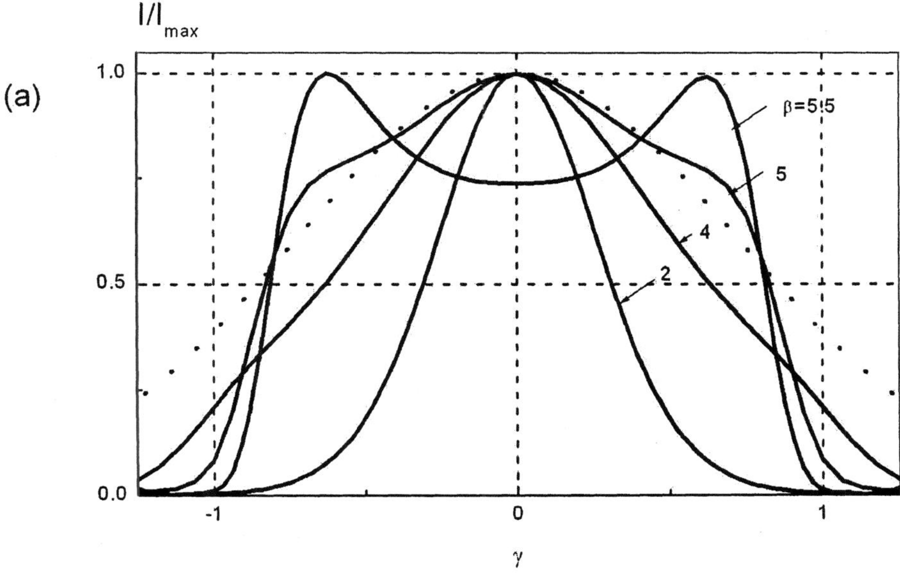

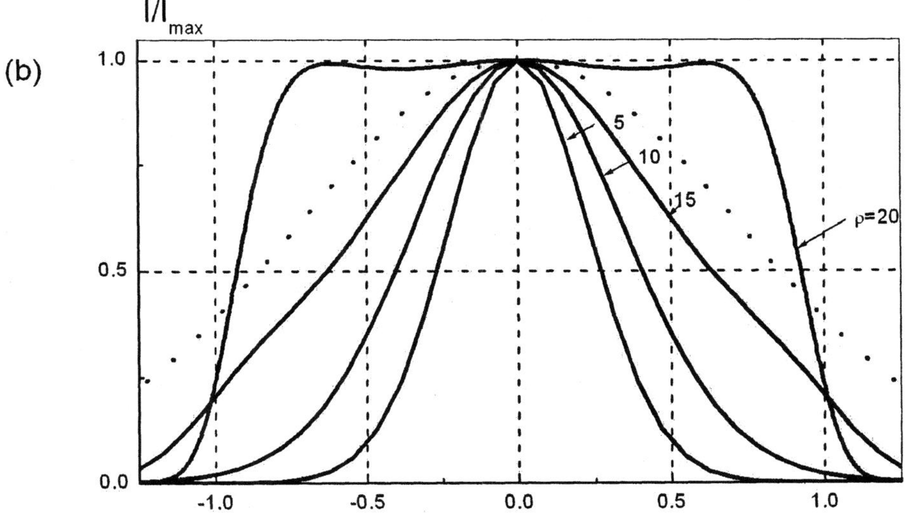

Figure 6: Averaged laser spectra per the first spectral dynamics cycle (calculated with the set of equations (18)): (a) for different β by ρ=15, and (b) for different $\rho=W/W_0$ by β=4 (dot-line is the dispersive resonator Sp.-Func. with its half-width σ_0).

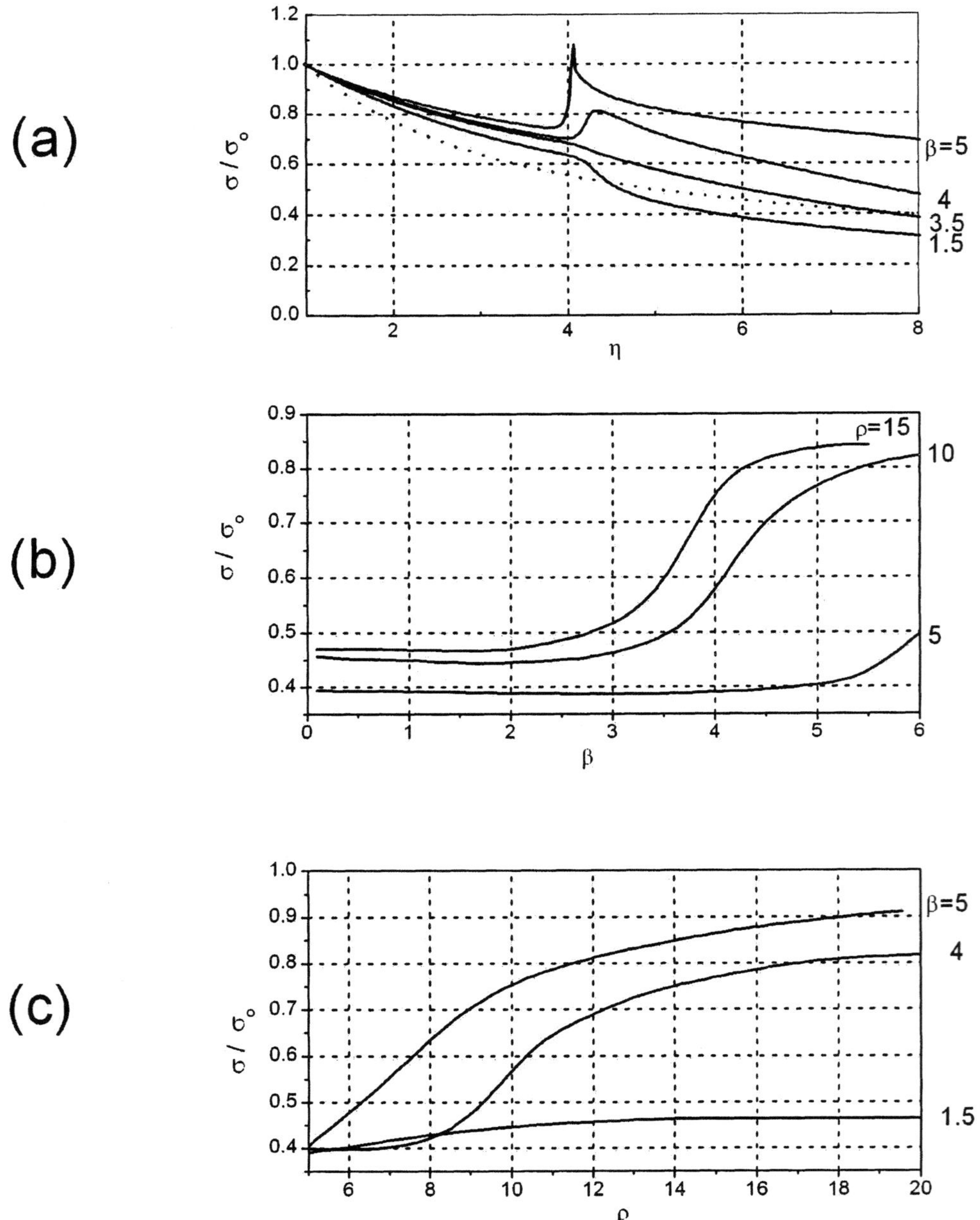

Figure 7: Influence of nonlinear dynamics on spectrum width: (a) dependence of its current value on time ($\eta=t/\tau_i$) for different aperture parameter β with fixed pumping energy ($\rho=W/W_0=15$), (b) dependence of its mean value per the first cycle of spectral dynamics on β for different ρ, (c) on ρ for different β (dot-line is the dispersive resonator Sp.-Func. with its half-width σ_0).

the following. During the linear stage of oscillation development (the pass number is equal to $\eta = t/\tau_l \leq 3$) there is a gradual narrowing of the spectrum, wich is accompanied by partial voiding of the peripheral regions of the excitation zone in the active medium. At the nonlinear stage there is diminution of the amplification constant in the zone center. If the size of the peripheral region is sufficient for arrangement of the new mode possessing greater diffraction losses (it is implemented at $\beta \geq 3.5 \div 4$), and if the pumping energy is enough for exceeding the amplification constant of the peripheral region over the "saturated" amplification constant of the center region, the spectrum starts to deform because of development of oscillation in the peripheral region ($\eta = 4$), being transformed in double-peak spectrum ($\eta = 6$). Then the "saturated" amplification constant of the peripheral region becomes less then the linearizing amplification in the center. At this stage ($\eta = 7 \div 10$) the inverse dynamics of spectrum transformation to the single-peak shape is implemented.

Fig. 5b corresponds to some dynamic balance between processes in the central and peripheral regions that practically gives rise to the invariance of the spectrum shape and its width at the nonlinear stage of oscillation.

All results illustrated by Fig. 5,6,7 are related to the first cycle of the spectrum nonlinear dynamics. The analysis, carried out by us, has shown that these processes can arise repeatedly if laser pulse duration is increased, but they run differently due to differences in starting condition. Simultaneously the character of processes of the first cycle depends on the choice of pumping function, on interrelations between the linear stage duration and that of nonlinear stage etc. In many respects it will be also defined by the presence of non-stationary inhomogeneities in the active medium (by variation of the excitation zone, thermal lenses in it etc.). Therefore we restricted ourselves to indicating the most general effects.

6　Spectral Dynamics in the Laser with Synthesis of the Spectrum Shape.

The main purpose of the present section is to formulate corrections for the spectrum tuning algorithm in accordance with contribution of the processes in active medium. To do this we shall analyze the spectrum kinetics in a laser with the composite Sp.-Func. of the dispersive cavity. Let us note that the direct solution of the complete set of the equations such as (5) or (18) in present case seems little informative and difficult, if possible at all, because it is necessary essentially to magnify the number of modes taking into consideration.

It is convenient beforehand to group together photon numbers related to modes of different partial resonators by introducing labels q_{rj}, where r is the number of the partial resonator. In this case the laser equations look like:

$$\frac{dN_2(x,t)}{dt} = W_p(N_T - N_2(x,t)) - U \cdot N_2(x,t)\sum_r\sum_j Z_{rj}^N(x)q_{rj}(t) - \frac{N_2(x,t)}{\tau}$$

$$\frac{dq_{rj}(t)}{dt} = q_{rj}(t)\cdot(V_a U \cdot \int_{-\infty}^{+\infty} Z_{rj}^q(x)N_2(x)dx - \frac{\gamma+\gamma_{rj}}{\tau_l})$$

Furthermore, for consideration of interaction of fields of different partial resonators in the same active medium, we shall characterize their spatial structure by distributions of central longitudinal modes, and assume that $Z_{rj}(x)=Z_r(x)$. Then

$$\frac{dN_2(x,t)}{dt} = W_p(N_T - N_2(x,t)) - U \cdot N_2(x,t)\sum_r Z_r^N(x)q_r(t) - \frac{N_2(x,t)}{\tau};$$

$$\tag{20}$$

$$\frac{dq_r(t)}{dt} = q_r(t)\cdot(V_a U \cdot \int_{-\infty}^{+\infty} Z_r^q(x)N_2(x)dx - (\gamma + \sum_j \gamma_{rj}\Phi_r(\lambda_j,t))/\tau_l),$$

where $\quad q_r(t) = \sum_j q_{rj}(t), \qquad\qquad\qquad\qquad \sum_j \Phi(\lambda_j,t) = 1,$

$\Phi_r(\lambda_j,t)$ defines temporal changes of the partial resonator spectra. Thus equations (19) describe the major effects of influence of distinctions of spatial distributions of mode fields in different partial resonators, which prevail in the present case. The evolution of spectra of partial resonators and their fields can approximately be taken into account in the same way as in the previous section, from the solutions of the equation set (18).

Now we reformulate the spectrum tuning algorithm. Using Sp.-Func. of the dispersive cavity for determination of the loss function, one can find, in principle, solution of the equation set (19) and present it as

$$Q(\lambda,t)=Q_n(\lambda,t)\,I(t)$$

where $Q_n(\lambda,t)$ is the normalized part of the solution ($\int Q_n(\lambda,t)\,d\lambda=1$) which characterizes the spectrum shape change, and $I(t)$ is the factor describing the change of the radiation energy. Taking into account that these functions obey the generalized theorem of the mean, it is possible to present the integrated intensity observable experimentally as

$$Q_T(\lambda) = \int_0^T dt\,Q_n(\lambda,t)I(t) = Q_n(\lambda,t_s(\lambda))\int_0^T dt\,I(t),$$

where T is the generation pulse duration, $t_s(\lambda)$ are some fixed time values satisfying the requirement $0<t_s(\lambda)<T$.

Now we modify the algorithm introduced in section 4 considering it as the algorithm of synthesis of the function $Q_n(\lambda, t_s(\lambda))$.

In the first step widths of Sp.-Func.'s of partial resonators are replaced by mean widths of spectra which can be found from solution of the equation set (18) obtained in Section 5. This corresponds to introduction of extra factors (< 1) before σ in the equations (15-17). As a result of their solution, one finds new values of $a_j(j=-N,\ldots\ldots,N)$ and Δ (these are changed in the direction of increasing the role of the spatial heterogeneous broadening).

In the second step we shall take into account that the found interrelation of amplitudes a_j corresponds not to the starting spectrum function, but to obtained as a result of its evolution in the laser (at $t_s(\lambda)=T_s$ this function really would coincide with the value of current spectrum $Q_n(\lambda, T_s)$ of laser for $t=T_s$). Therefore in experiment such spectrum will be realized at smaller differences between values a_j and some reference a_n (for example, peak), which can be found from the analysis of the laser equations. The correction factors h_{jn} should be equal to the ratio of differences of intensities for mean wavelengths of different partial resonators in the spectrum $Q_n(\lambda, t_s(\lambda))$ to that in the starting spectrum corresponding to $Q_n(\lambda, t_s(\lambda))$.

The greatest values $^h h_{jn}$ of these coefficients will occur in the laser with the homogeneous broadening of the spectrum. In this case, when the multiplicative transformation of a spectrum is implemented, the values of coefficients $^h h_{jn}$ can be found precisely from radical relations or evaluated in linear approach (frequently used in the intracavity laser spectroscopy) from the expression $^h h=1/\eta_s$, where $\eta_s=T_s/\tau_l$ is the number of passes corresponding to T_s.

The presence of the spatial heterogeneous broadening leads to the different decrease of amplification constants of different partial resonators at the expense of nonlinear effects (the amplification constant is decreased in the greater degree for the resonator with smaller losses). It is possible to improve coefficients h_{jn} by the iterative solutions of the set (20) for synthesis of a specific spectrum or by introducing a common correction multiplier h_n for the laser with the given kinetics (under our condition this more often leads to the inverse effect, namely to decrease of coefficients h_{jn} in comparison with $^h h_{jn}$.). Because the model can not take into account all factors influencing a spectrum in the actual laser, in the final stage one can more precisely tune the necessary spectrum shape just by watching it visually on the computer's monitor.

In the following section we shall illustrate the application of such algorithm to synthesis and tuning of actual experimental spectra (Fig.11, 12). (Partially problems of sections 5, 6 are considered in [33]).

7 Experiment

This section contains the results of experimental studies of a proposed setup (Fig.1a,8) implementing the laser system with the emission of continuous spectra, which general properties were discussed in Sec.4-6.

As noted above, a key step in practical realization of the control element 2 is the elaboration of an efficient tuning method (and above all the electronic one). A flexible control over two spatial distributions, P-Dist. and N-Dist, should be ensured in our case. Because the requirements for the first are not very stringent, one can use several alternative methods. In our case, we created a parabolic P-Dist. by slightly detuning the telescope 4 (Fig.1a, 8). If needed, the distance between lenses can be tuned quickly by the electrooptic or piezoelectric tools.

N-Dist. must have essentially more complex and flexible tunable structure. Acoustooptic units 3 working in the spatial modulator regime offer a wide variety of features that can be used for this purpose. The regime is especially convenient for the lasers in study because of essential distinction between the laser pulse duration (~ 10 ns) and the time needed to fill the acoustooptic modulator 3 with a traveling sound wave (typically $>1\mu s$). Under these conditions the light sees the sound wave as a motionless diaphragm. Its shape (N-Dist.) can be quickly synthesized and tuned by the choice of the shape of the temporal amplitude modulation of the microwave signal exciting the piezotransducer of El.3. In a traveling sound wave this temporal modulation is converted into the modulation of spatial profile along the direction of its propagation. Evidently for this regime to be realized, it is necessary to match the modulation signal duration T_p (see Fig.8) with the acoustooptic unit's length and to synchronize the moment of the laser pulse excitation with the necessary sound wave position.

The proposed experimental setup (Fig.8) is built around a short pulse dye (Rhodamin 6G) laser ($t \sim 10$ ns). A quasi longitudinal pumping of the cell 6 was achieved with the second harmonic of YAG:Nd^{3+} laser 10 (LTI-403, $\lambda = 530$ nm, $\tau \sim 14$ ns). Spectrum analysis was performed by a linear CCD array 8 mounted on spectrograph (UF-90) 9 into which the beam reflected from a face of prism expander 5 was introduced.

The lenses of telescope 4 have had the focal lengths 35 mm and 15 mm. The intracavity spatial acoustooptical modulator 3 has been made of TeO$_2$ crystal with aperture 20 mm x 5 mm and has had an optimal geometry of the acoustooptical interaction[31] (i.e. the sound wave is excited in the plane [110], the angle between the transducer surface and the crystal optical axis is approximately equal to $6°$).

The electronic part of the setup includes the microwave generator 11 (G4-116) which output has been modulated from a home built interface 12 with the fast digital-analogue converter. The modulation signal shape could be synthesized by computer 14 (IBM PC AT). A second home built interface 13 was used to synchronize the modulation pulse with the laser impulse.

The possibility of formation and control of the spectrum structure in our laser is illustrated in Fig.9-12, which comprise the most significant experimental results. Fig.9a demonstrates the gradual transition between the regimes of similar and non-similar spectrum conversion for example of four-frequency spectrum transformation with change of Δ keeping the other parameters constant. As noted above, these regimes are due to different dependencies on Δ of the distance between spectral lines and their widths.

Fig.10-12 shows the ability to control the shape of continuous spectrum. As controlled parameters, we have chosen the following functions of the moments of the spectral distribution $I(\lambda)$ [27]: the dispersion $\sigma=(M_2)^{1/2}$; the asymmetry $\gamma_1 = M_3/\sigma^3$, the kurtosis $\gamma = M_4/\sigma^4 - 3$, where $M_k = \int(\lambda-m_1)I(\lambda)\,d\lambda$ is the central moment of order k, and m_1 characterizes the mean wavelength (note that traditional tuning of the wavelength corresponds to the change of m_1 only).

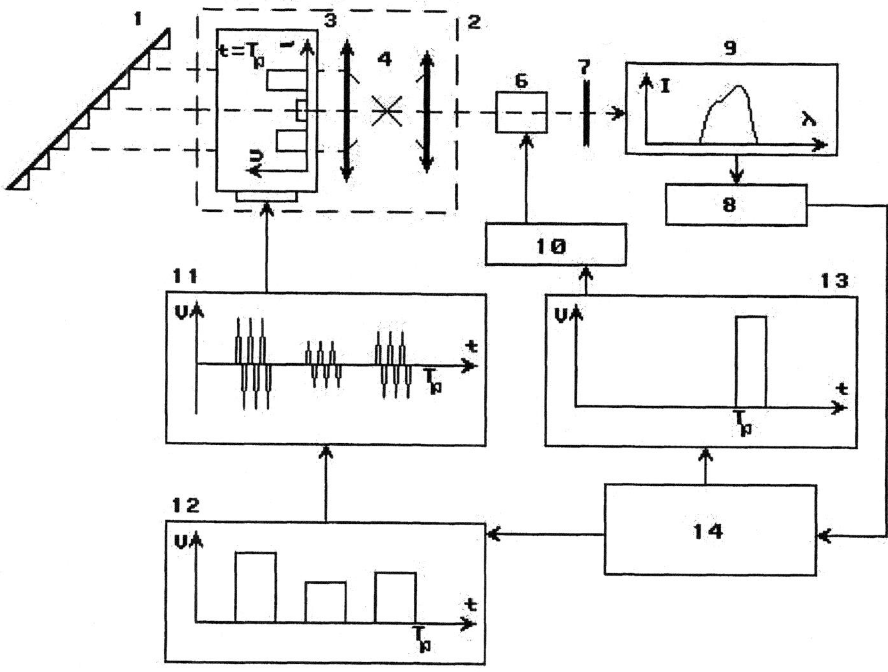

Figure 8: Experimental setup with the laser having the tunable shape of the continuous spectrum and signals in its different units. Here are grating 1, control element 2, acoustooptical spatial modulator 3, lens telescope 4, cell with the dye solution 6, mirror 7, CCD array photodetector 8, spectrograph 9, laser of pumping 10, microwave generator 11, interfaces 12 and 13, computer 14 (the prism expander 5 shown in Fig. 1 is missed for simplicity).

The change of the spectrum width can be regarded as a first simplest step in the spectrum formation. Similar task was already considered in [17,18,20,32] where the width adjustment was achieved with the intracavity acoustooptical filter [32], the aperture stop of variable size [17], the prism or lens expander [18,20]. In the laser in study we can demonstrate the control over the spectrum width by changing either Δ (Fig.9a) or the widths of the intracavity diaphragm (Fig.9b). In the latter case the measurement were made with the sound pulses having different duration and the same amplitude.

Moreover, our method of laser spectrum synthesis is capable to ensure much more control over the shape of the laser emission spectrum. Fig.10, 11 show the results of spectrum shape synthesis where the tunable parameter was the third moment of the spectrum which defines its asymmetry. In these experiments, the signal of the sound wave modulation was created in the form of either discrete pulses (Fig.11, where, without loss of generality, only the case of three diaphragms is shown) or continuous distribution (Fig.10). Fig.12 shows the control of the spectrum kurtosis that characterizes either a gap or peak sharpness in the middle.

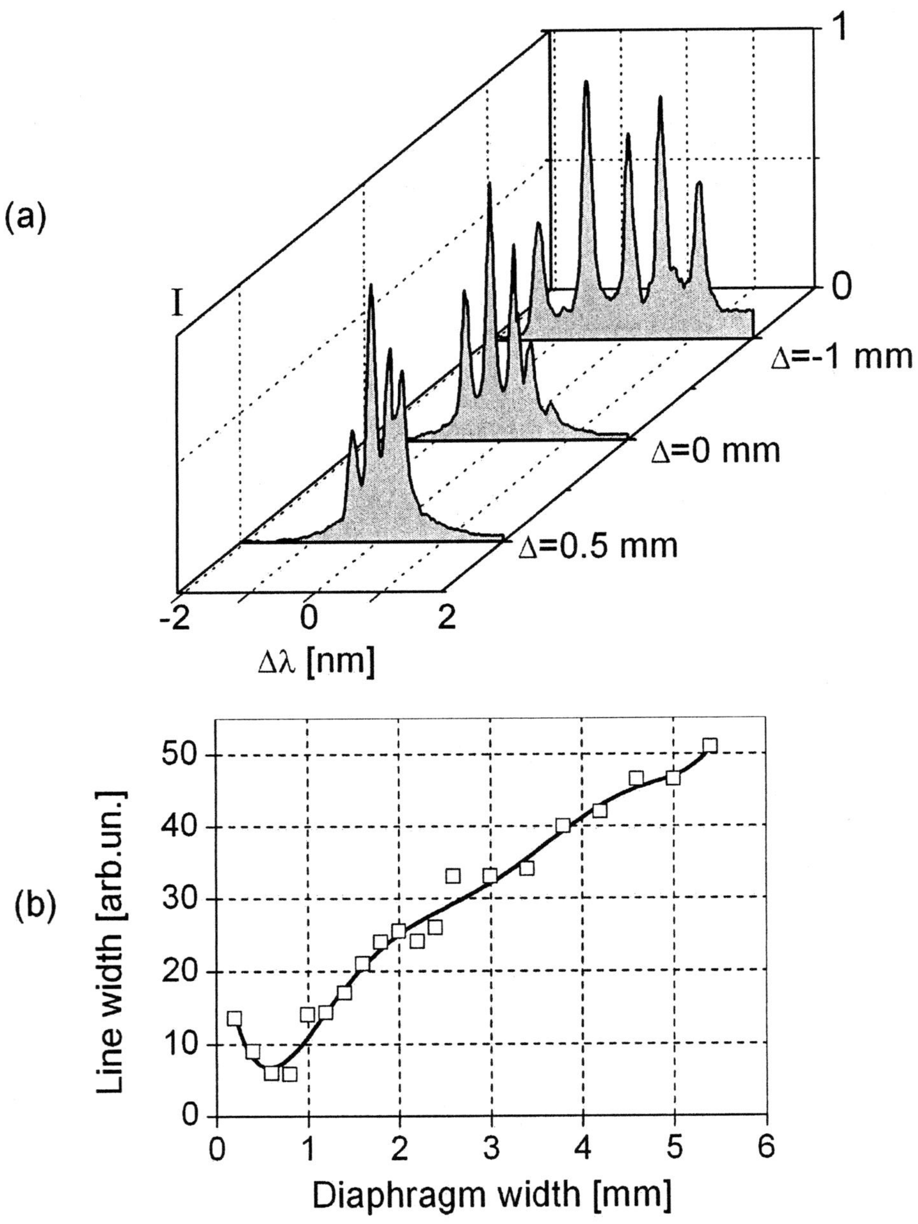

Figure 9: Experimental dependence (a) of laser spectrum on the telescope detuning Δ, (b) of the laser spectrum width on the intracavity diaphragm size (the sound wave pulse duration).

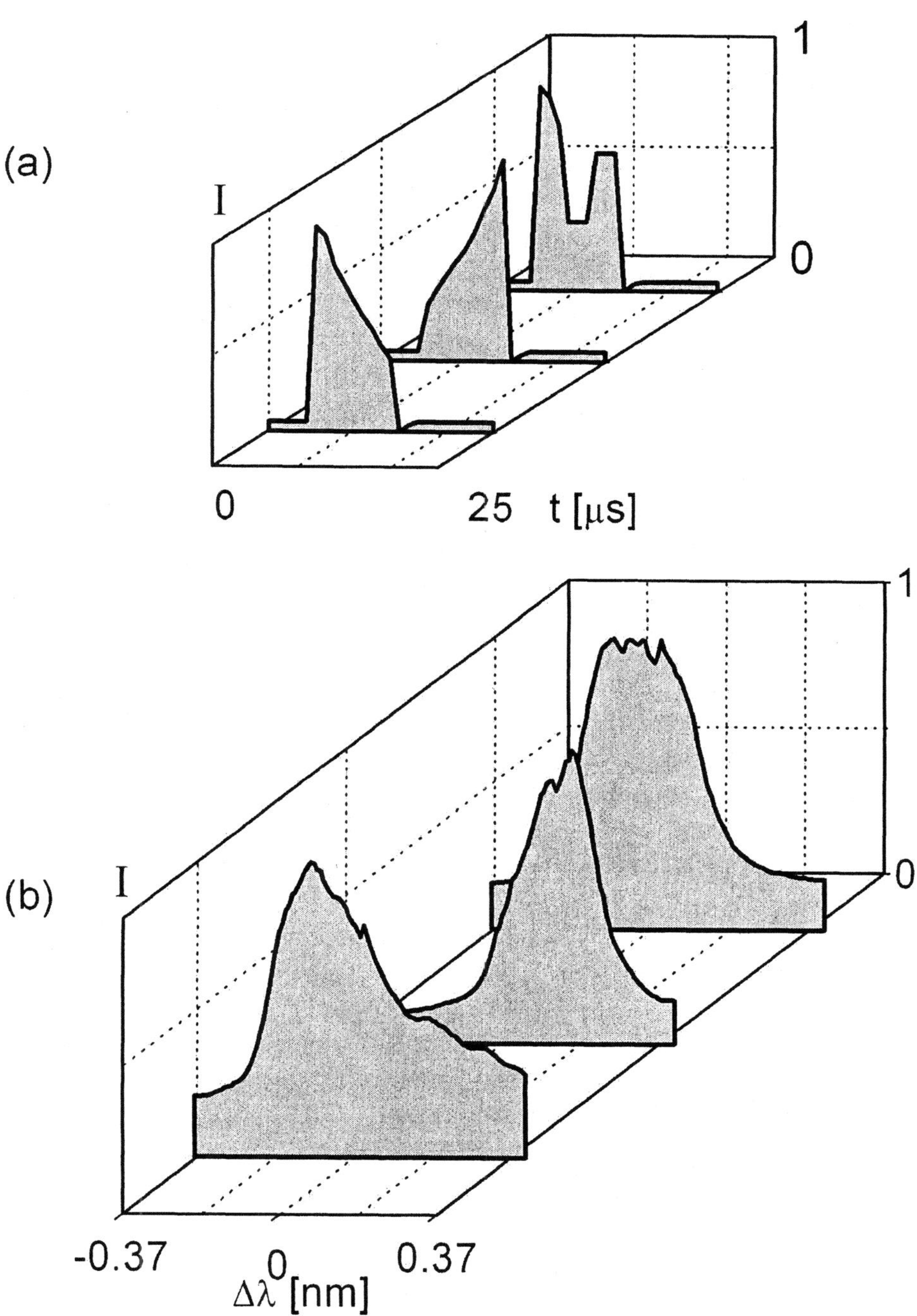

Figure 10: Spectrum asymmetry control (modulation with continuous signal). Relation between sound wave shapes (a) and experimental laser spectra (b).

Fig. 11c, 12c show also the calculated spectra for $\eta = T/\tau_l \approx 4$ (corresponds to our laser parameters), which are obtained with the formulated above algorithm of synthesis and illustrate the satisfactory agreement with experiment.

The visible distinction of the laser spectrum from the modulation signal is determined both by the nonlinearities inherent to the non-similar spectrum transformation, and by technical causes (such as the influence of the profile of the excitation zone in the active medium, inhomogeneouties both of the sound and of the light beam that fills the

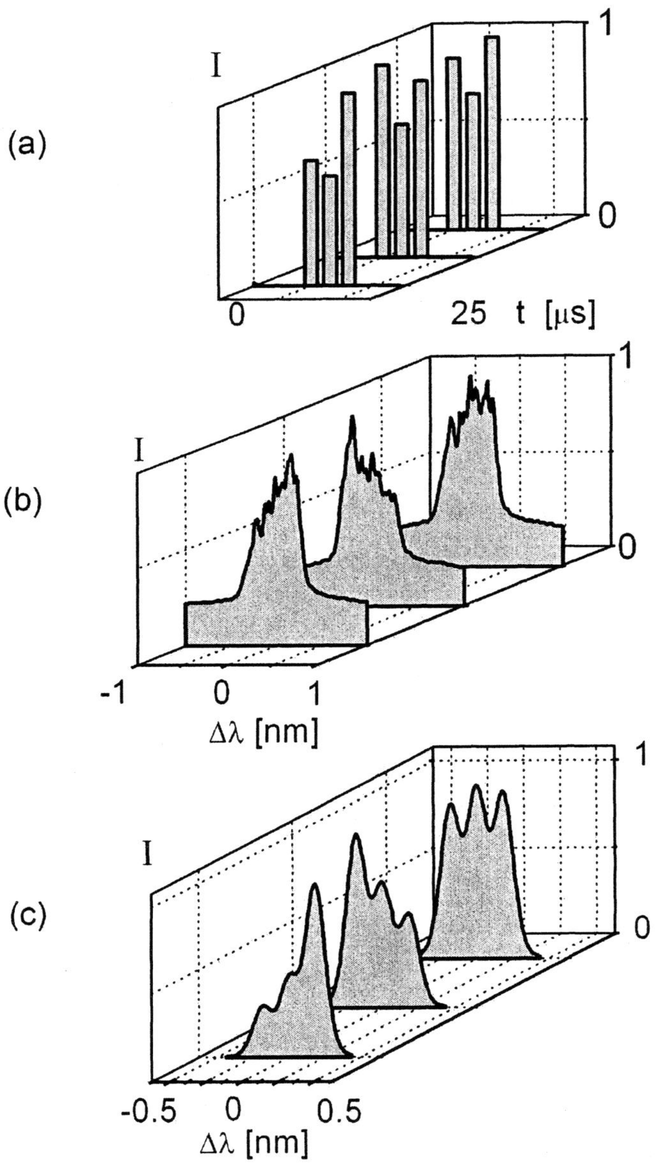

Figure 11: Spectrum asymmetry control (modulation with discrete pulses). Relation between sound wave shapes (a), laser experimental spectra (b), and theoretical that (c).

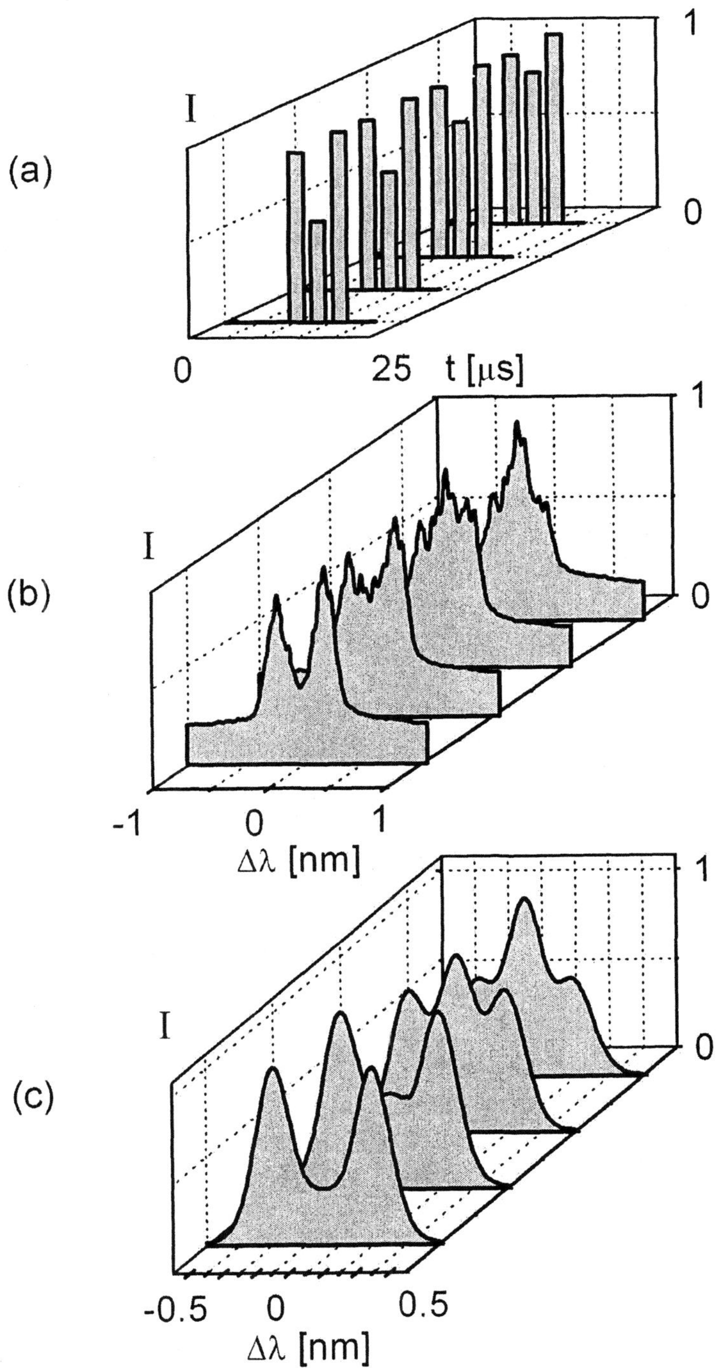

Figure 12: Spectrum kurtosis control. Relation between sound wave shapes (a), laser experimental spectra (b), and theoretical that (c).

modulator aperture etc.). At the same time these differences confirm the presence of the spatial heterogeneous broadening (different influence of borders of the excitation zone in the active medium that depends on a specific alignment of the resonator units).

The visually different contribution of the central partial resonator and the boundary ones into the corresponding spectrum regions is defined by the mentioned above factors and by different contribution of edges of spectral functions of different partial resonators in central and boundary points. The contribution of the last factor is even more pronounced for rectangular sound impulses, the results for which are given on experimental dependencies of Fig. 11,12.

To eliminate the influence of technical factors, the sound distribution for concrete realization of laser should be beforehand calibrated for symmetric spectra at the linear region or for the pulse with the more short duration.

Finally, it is worth noticing that the proposed method allows not only to tune several spectrum moments but to construct spectra with more complicated structure, even the discrete ones (Fig.9a).

8 Conclusions

Among the results of the present work we would like to highlight the following:

(1)   The regularities of the evolution of the Sp.-Func. of two diffractively coupled resonators when β increases in intermediate region ($\beta\geq1$). First, Sp.-Func. has only one peak independently of resonator configuration (Re g). Then, three maxima appear. At last, only two maxima exist, and their position can be smoothly tuned with the change of $|g-1|$ and can be calculated using the obtained approximate relation.

(2)   The regularities of the dynamics of a single spectral line in lasers with resonators containing units with the angular dispersion. In the intermediate range of the apertures with $\beta\geq4$, the spectrum can transform its shape from that consisting of one-peak to double-peak and back through the intermediate phases. The equilibrium and practically constant spectrum shape can be observed in the nonlinear stage at a particular pumping energy and its pulse shape. The spectrum kinetics is familiar with its gradual narrowing below this pumping energy.

(3)   The model defining the synthesis of Sp.-Func., when controlled parameters are the spatial distribution of the amplitude norm and phase of the transmittance in cross-section (N-Dist. and P-Dist.), and the equation set for finding them in accordance with the given spectral moments. Algorithm of analog synthesis of the spectrum shape which takes into account, in parallel with features of selection of waves in dispersive cavities, the influence on the spectrum shape of processes of a competition in the laser and nonlinear effects arising due to "the spatial heterogeneous broadening" in the active medium.

(4)   The realization and experimental performances of the laser with the tunable radiation spectrum shape that has continuous distribution. An efficient control of the spectrum shape is achieved with new control device that includes the acoustooptic spatial modulator and the lens telescope with the variable distance between its lenses.

Our results are not limited to a specific problem of the continuous spectrum synthesis. In particular, we hope, it will be yet another step toward both the development of the dispersive resonator inverse task solution methods and understanding of the selection processes in such systems (and above all in the coupled dispersive resonators).

The obtained by us regularities of influence of the active medium nonlinearity on spectrum dynamics also can have more general importance. We have considered simple enough but practically important case, however, a set of effects due to it can be much more plentiful. If the duration of a cycle of nonlinear dynamics is increased, the reverse reaction of nonlinear effects on spatial mode structure should necessarily be taken into account. They can be analyzed using system (5) with modes of higher order. The similar situation will take place at $\beta > 10$ when the differences in losses for modes of the different order is much less. The character of similar effects will be also influenced by non-stationary changes in active medium parameters (for example, induction of thermal lenses, transverse displacements of an excitation zone because of variations of the pumping laser performances, etc.), by interference effects in the active medium (for example, in crystals of the corundum with an titan ions impurity) which can change the Sp.-Func. shape, etc.

At the same time, as a starting point for our analysis we could choose the resonator including a lens [16] instead of the flat-mirror configuration. This choice may influence the specific details of the laser setup, but the final results will not have to depend on it because the laser must operate in the non-similarity regime and must have the intermediate configuration.

The proposed specific implementation of the control element is not of principal importance. For example, both procedures (setting N-Dist. and P-Dist.) may be performed by just one acoustooptical element if the frequency modulation of microwave signal is used in combination with the amplitude one.

At last, it is necessary to note that problems of spectra synthesis belong to the inverse problems of laser physics. Because of their complexity it is actually impossible to solve them by only mathematical methods. Their solution is usually founded on the use of more simple models following from the peculiar features of the processes in such systems. Therefore the study of such problems is significant not only practically, but also is important from the scientific point of view for further and more thorough development of physics of tunable laser.

References

[1]   V.I.Kravchenko, Yu.N.Parkhomenko, "Electronic tuning lasers on condensed media," Izvestija AN SSSR,Ser.Phys. **54**, 1543-1551 (1990), (in Russian).

[2] K.O.Nvairo, I.H.White, C.J.Armistead, P.A.Kirkby, "Multiple channel signal generation using multichannal grating cavity laser with crosstalk compensation," Electronics Letters **28**, 261-263 (1992).

[3] J.B.D.Soole, K.Poguntke, A.Scherer, H.P.LeBlanc, C.Chang - Hasnain, J.R.Hayes, C.Caneau, R.Bhat, M.A.Koza, "Multistripe array grating integrated cavity (MAGIC) laser: a new semiconductor laser for WDM applications," Electronics Letters **28**, 1806-1807 (1992).

[4] I.H.White, "A multichannel grating cavity laser for wavelength division multiplexing applications," IEEE J.Lightwave Technology **9**, 893-899 (1991).

[5] M.Alden, K.Fredricksson, S.Wallin, "Application of a two-color dye laser in CARS experiments for fast determination of temperatures," Appl. Opt. **23**, 2053-2055 (1984).

[6] M.Namazian, R.L.Schmitt, M.B.Long, "Two-wavelength single laser CH and CH_4 imaging in a lifted turbulent diffusion flame," Appl. Opt. **27**, 3597-3600 (1988).

[7] A.Arnold, W.Ketterle, H.Becker, J.Wolfrum, "Simultaneous single-shot imaging of OH and O using 2-wavelength eximer laser," Appl. Phys. B **51**, 99-102 (1992).

[8] D.Bruneau, H.Cazeneuve, C.Loth, J.Pelon, "Double-pulse dual-wavelength alexandrite laser for atmospheric water vapor measurement," Appl. Opt. **30**, 3930-3937 (1991).

[9] N.N.Nenchev, M.M.Martin, Y.A.Meyer, "Alternative wavelength DIAL dye laser using a reflecting interference wedge," Appl. Opt. **24**, 1957-1959 (1985).

[10] H.V.Pitingsrud, "CO laser for lidar applications, producing two narrowly spaced independently wavelength-selectable Q-switched output pulses," Appl. Opt. **30**, 3952-3963 (1991).

[11] H.Takeda, Y.Akabane, F.Kannari, "Dual-wavelength operation of a flashlamp pumped narrow-linewidth Ti:Sapphire laser," Jpn. J. Appl.Phys. **33**, 6557-5663 (1994).

[12] Yu.N.Parkhomenko, O.N.Galkin, O.V.Gorbenko, V.A.Sokolov, "Laser with rapid tuning of the spectrum shape," Applied Optics **37**,   (1998).

[13] V.I.Kravchenko, A.I.Lyushchenko, Yu.N. Parkhomenko, "The competition of spectral components of a tunable laser accompanying the biharmonic excitation of an acoustooptic deflector," Sov. J. Quantum Electronics **22**, 40-43 (1992).

[14] O.N.Galkin, Yu.S.Plaksij, Yu.N.Parkhomenko, "Formation of a multifrequency spectrum of a tunable laser by excitation of spatially modulated gratings in acoustooptic deflector," Quantum Electronics **25**, 127-129 (1995).

[15] O.N.Galkin, V.I.Kravchenko, A.I.Liushenko, Yu.N. Parkhomenko, "Light diffraction from multifrequency volume gratings in anisotropic media," International J. Nonlinear Opt. Phys. **3**, 55-68 (1994).

[16] V.I.Kravchenko, Yu.N.Parkhomenko, "A calculation of the selectivity of resonators with the angular dispersion," Quantum Electronics **12**, 1220-1226 (1985), (in Russian).

[17] V.I.Voronzov, V.I.Kravchenko, Yu.D.Opanasyuk, Yu.N.Parkhomenko, "Selective properties of a linear resonator with a nonsymmetrical dispersive element," Quantum Electronics **9**, 1586-91 (1982), (in Russian).

[18] V.I.Kravchenko, Yu.N.Parkhomenko, V.A.Sokolov, "The selectivity of dispersive resonators with arbitrary placed focusing and bounding elements," Quantum Electronics **13**, 2038-43 (1986), (in Russian).

[19] V.I.Kravchenko, Yu.N.Parkhomenko, V.A. Sokolov, "Selectivity of a dispersive resonator with an inhomogeneous aperture," Sov. J. Quantum Electronics **18**, 54-57 (1988).

[20] V.I.Kravchenko, Yu.N.Parkhomenko, V.A.Sokolov, "Effect of a dynamic lens on the spectral characteristics of a pulse organic duy laser with the dispersion cavity," Optics and Spectroscopy **67**, 550-554 (1989).

[21] V.I.Kravchenko, Yu.N.Parkhomenko, "Selectivity of dispersive cavities with random spatial inhomogeneities," Izvestija vuzov, Ser.Radiofiz. **37**, 330-339 (1994). In Russian.

[22] V.I.Voronzov, V.I.Kravchenko, Yu.N.Parkhomenko, "Ring dispersive resonators with grating," Pisma v JTP **6**, 1105-1109 (1980).

[23] V.I.Kravchenko, Yu.N.Parkhomenko, V.A. Sokolov, "The selecttivity of coupled dispersive resonators," Sov. J. Quantum Electronics **18**, 1135-1139 (1988).

[24] Yu.N.Parkhomenko, "Role of aberration effects in dispersive cavities," Quantum Electronics **28**, 1102-1107 (1998).

[25] P.Flamant, "Spectral narrowing in pulsed dye lasers," Applied Optics **17**, 955-959 (1978).

[26] Ja.I.Hanin, *Quantum radiophysics. Dynamics of quantum generators*, V.2, (Moskow, "Sovetskoe Radio", 1975), In Russian.

[27] H.Cramer, *Mathematical methods of statistics* (Moskow, "Mir", 1975), (in Russian).

[28] E. Armandillo, G.Giuliani, "Achieveemnt of large sized mode from an eximer laser (apoditic filter)," Optics Letters **11**, 445-447 (1985).

[29] K.J.Snell, N.McCarthy, M.Piche, P.Lavigne, "Single transverse mode oscillation from an unstable resonator Nd:YAG using a variable reflectivity mirror," Optics Commun. **65**, 377-382 (1988).

[30] A.Parent, P.Lavigne, "Variable reflectivity unstable resonators for coherent laser radar emmiters," Applied Optics **28**, 901-903 (1989).

[31] T.Yano, M.Kawabushi, A.Fukumoto, A.Watanabe, "Anisotropic Light Deflector Without Midband Degeneracy," Appl. Phys. Lett. **26**, 689 (1975).

[32] M.F.Stelmah, V.G.Dmitriev, L.K.Mihajlov, S.L.Seregin, E.M.Spizin, O.B.Cherednichenko, "Tunable lasers and laser spectral devices with the use of acoutooptic filters," J. Appl. Spectroscopy **40**, 181-189 (1984), (in Russian).

[33] Yu.N.Parhomenko, O.V.Anisimova, "Influence of spatial distribution of modes on spectrum evolution in tunable lasers," Quantum Electronics **31**, 727-732 (2001).

Time Resolved Studies on Single Pulse Characteristics of a Copper Vapor Laser with Different Resonators and Their Impact on Second Harmonic Generation and Dye Laser Pumping

Om Prakash, S. K. Dixit, H. S. Vora, N. M. Shenoy & R. Bhatnagar
Laser Systems Engineering Division
Centre For Advanced Techonology
Indore- 452013
INDIA

ABSTRACT

Beam quality of a self-terminating laser like copper vapor laser (CVL) has been an issue of concern since its advent. Recent advances in resonator design coupled with time resolved study of the beam characteristics has led to a reassessment of the potential of these lasers in applications. In this paper, results of time averaged and time resolved studies on the development of the CVL characteristics for different resonators are presented. It is shown that a generalised diffraction filtered resonator (GDFR) is best suited for CVL as compared to unstable resonators (UR) of magnification as high as 500. The effect of these beams of varying spatio-temporal characteristics, on two representative applications namely second harmonic generation and transverse pumping of dye laser were studied. The frequency conversion efficiencies were found to be highest for GDFR as compared to UR for same fundamental beam power. In case of dye laser pumping it is shown that the bandwidth and the divergence depend on the coherence width of the pump beam. Higher the coherence width of pump beam lower is the divergence and the bandwidth of the dye laser.

Introduction

The copper vapour laser (CVL), because of high average power and high repetition rate operation, has found use in many applications like nonlinear frequency conversion

[1], material processing [2] and dye laser pumping [3]. However its full potential could not be exploited due to poor output divergence/coherence. Recent advances in resonator design for pulsed high gain lasers [4] and its application to CVL [5] has led to diffraction limited output, neccessiating a re-assessment of the current understanding of CVL beam quality and its impact on applications. It is known that the average characteristics of a pulsed beam depends on the evolution of the characteristics within the pulse. It is the time-resolved study of various parameters that give insight into the behavior of the characteristics and its effect in applications. With this in mind, two applications, namely second harmonic generation and transverse pumping of dye laser were chosen for the study. In SHG, where divergence or the coherence of the interacting beams is important, it is shown that intrapulse variation of divergence plays a decisive role in efficiency, divergence and amplitude stability of the SHG beam. On the other hand, in case of transverse pumped dye laser, pump beam divergence is not expected to play significant role in deciding the characteristics of the dye laser. It is shown in this case also the divergence/coherence of the CVL beam influences the divergence and the bandwidth of the dye laser.

The first section deals with study of the time-averaged and time-resolved spatial coherence/divergence of CVL with generalized diffraction filtered resonator (GDFR) and unstable resonators of magnifications varying from 12.5 to 500. In the second section we discuss, through time resolved studies, how beams of different coherence properties affect the conversion efficiency of the second harmonic generation and the output characteristics of transverse dye laser.

1 Beam quality issues in copper vapor laser

The coherence properties of CVL attracted early attention from Hargrove et al [6] who demonstrated, using a streak camera, that in an unstable resonator the divergence varies in time reaching diffraction limit only towards the end of the pulse. Several studies, using unstable resonators of various magnifications, evaluated the time averaged divergence of the output beams [7-9]. Magnification of 12.5 to 200 were used. The results indicated that the divergence reduces as the magnification is increased. The time resolved coherence studies using reversal shear interferometer indicated that the coherence width increases after each transit as the radiation traverses the resonator [10-11]. The divergence also follows this step like decrease by a factor M after each transit [12]. These studies were conducted using laser tubes of differing cross-section varying from 25 mm to 40 mm and varying length from 80 cm to 120 cm [12-14]. It is known that the gain in CVL is not uniform and has a radial profile which depends on discharge tube cross-section and length in addition to other operating parameters [14]. With difference in radial gain profile for different tube diameters and length, it is difficult to compare the results of different authors on the beam characteristics and analyze the role of the resonator. We present a comparative study on time averaged and time resolved characteristics of the output beams from unstable resonators of magnification upto 500 and diffraction filtered resonators of equivalent magnifications on same discharge tube under identical operating conditions.

Experimental set-up

The laser used in the experiment has a discharge tube of 28 mm diameter and gain length of about 100 cm (Fig. 1a). The total output power with a plane-plane resonator was 18W (5.5 kHz) with 13.6 W at 510.6 nm and 4.4 W at 578.2 nm wavelength. The resonator consists of mirrors M_1 and M_2, separated by cavity length L (Fig. 1a). The focal length $f_{1,2}$ of the mirrors $M_{1,2}$ can be selected for resonators of various types. In UR, magnification $|M_U| = f_1/f_2$ is varied by changing the focal length of M_2. Intra-cavity apertures of various diameters were placed, close to M_2, for GDFR of various magnifications. Table lists the resonators parameters and results of power and coherence measurements. In GDFR, radiation is diffracted by an intracavity aperture, P (dia. 2a), placed close to plane mirror M_2 and propagates into farfield as it travels to convex mirror M_1. The central lobe of farfield intensity distribution selected by another aperture is reflected back into the gain medium. The aperture diffracts the spherical wave originating from the virtual source S, which is the image of aperture P in M_1. The round trip magnification given by $M_G = 0.61x[(2+L/f_1)x\lambda xL/a^2]$ can be varied by varying aperture diameter, 2a [5].

A home made reversal shear interferometer [15], consisting of wedge beam splitters/combiners, (W_1, W_2) and mirrors, (M_3- M_7) for beam splitting and wavefront reversal was used in the measurements (fig. 1b). The incident beam ABC is split along two paths, OPQR and ODEFR and then recombined to provide wavefront reversal in one path due to an extra reflection. The two path lengths are made almost equal and the interference fringes are observed at W_2. The fringes were recorded by a CCD camera/frame grabber card and analysed using a software developed inhouse [16]. The coherence was estimated from the visibility of the fringes, V, given by

$$V = (I_{max} - I_{min}) / (I_{max} + I_{min})$$

The visibility of the fringes equal the degree of coherence if intensity in the two beams are equal. It is well known that in CVL, the coherence evolves in time with each transit of radiation through the resonator. The time averaged interferogram therefore will give lower limit of coherence, as superposition of interferograms for each transit reduces the visibility of the fringes. A reasonable measure of coherence is given by the coherence width, defined as the radial distance from the zero shear fringe to the fringe at which the visibility falls to 1/e of the visibility of zero shear fringe. The zero shear fringe is identified by locating the fringe with unit visibility. It is noticed that the visibility of the fringes does not reach unity anywhere across the beam. This is due to the presence of ASE, difference in intensity of the two interfering wavefronts and non-uniformity of the beam which may reduce its visibility to less than unity. Therefore the fringe with maximum visibility is located and designated as the fringe of zero shear and all radial distances are measured from its position. In all the measurements, care was taken to minimize the ASE by suitable spatial filters before the beam was allowed to fall on the interferometer.

Table. Resonator parameters and results

Types of resonators	f_1 (mm)	f_2 (mm)	L (mm)	2a (mm)	Magnification M	Output power (W) $P_{G\&U}$	$\frac{P_{GDFR,UR}}{P_T}$ X100	Coherence width(mm)
P-P	∞	∞	2300	-	-	13.6	-	not measurable
UR	2500	-200	2300	-	12.5	7.6	62	1.8
	2500	-75	2425	-	33	7.0	58	2.2
	2500	-25	2475	-	100	6.0	50	6.0
	2500	-10	2490	-	250	2.8	25	9.0
	2500	-7.5	2492.5	-	333	2.2	18	No 1/e fall
	2500	-5.0	2495	-	500	1.9	18	No 1/e fall
GDFR	-200	∞	2300	1.0	38	2.9	46	No 1/e fall
	-200	∞	2300	0.9	47	3.4	53	No 1/e fall
	-200	∞	2300	0.8	60	3.5	53	No 1/e fall
	-200	∞	2300	0.7	79	3.9	55	No 1/e fall
	-200	∞	2300	0.6	107	4.3	51	No 1/e fall
	-200	∞	2300	0.5	155	3.8	47	No 1/e fall
	-200	∞	2300	0.4	240	2.9	35	No 1/e fall
	-200	∞	2300	0.3	430	2.7	32	No 1/e fall

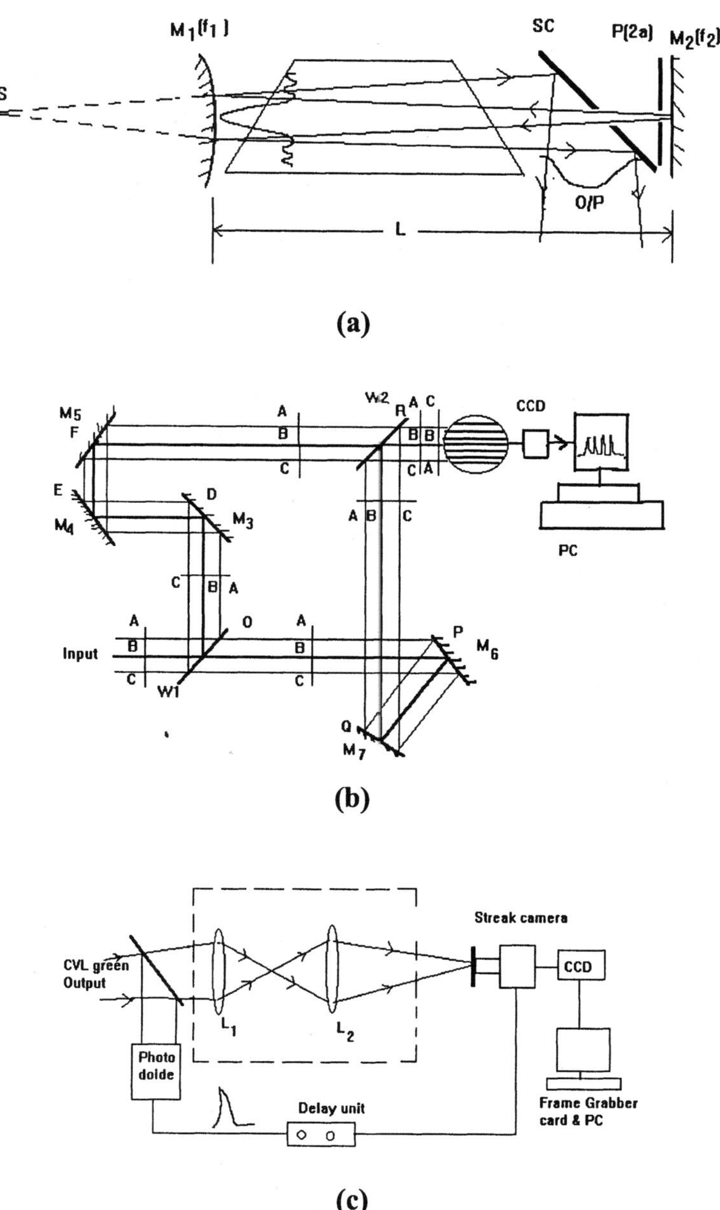

Figure 1: Experimental set-up for (a) CVL Resonator (b) Reversal shear interferometer (c) time resolved studies

(a). Average spatial coherence

Figure (2) shows the typical reversal shear interferograms of UR's (Fig. 2a-b) and GDFR's (Fig. 2c-d) of equivalent magnifications along with intensity scan across the diameter. The degree of coherence across the beam diameter is estimated from the fringe visibility data and is shown below the interferogram. In UR, (M_U=12.5) the fringe visibility was low and falls rapidly towards the edges. The coherence width increases from 1.8 mm to about 9 mm as M_U is increased from 12.5 to 250. For M_U of 333 and 500, the visibility did not fall to 1/e of the visibility of the zero shear fringe. In GDFR (M_G =38) the fringes have high contrast in the central region and decreases towards the outer region. As M_G is increased from 38 to 240 the high contrast region extends across the beam diameter indicating higher spatial coherence. It is also indicated by the degree of coherence curves shown below each figure. Since the visibility across the diameter does not fall to 1/e of the zero shear value it is difficult to assign a coherence width. Comparing performance of UR and GDFR for each magnification (Table), it is seen that the GDFR beam is coherent across its diameter for magnification as low as 38 whereas the unstable resonator beam somewhat matches the GDFR features only for M_U = 500. It is expected that the UR with high magnification should spatially filter the intracavity radiation in the same way as the diffracting aperture in the GDFR, i.e. by selecting a small part of the incident wavefront and launching into the resonator. The observed difference in coherence characteristics suggests different mechanism for the two cases. In GDFR, the radiation evolves from the diffraction of the single pass ASE right after the first transit. The radiation, diffracted from small aperture, goes to far field as it traverses to the convex mirror placed beyond the Rayleigh range and by selecting the central lobe of diffraction pattern, an uniphase smoothly varying beam is available for further transits. In unstable resonator, on the other hand, the round trip magnification does provide for small part of the wavefront to be reflected back to the gain medium in each successive transit but the radiation evolves from extremely small coherence width of the spontaneous emission. This difference in evolution can only be resolved by time resolved studies, the results of which are presented later.

The output power for all the resonator configurations were also measured (table). The output power, P_T, is composed of ASE and low divergence UR power (P_U) or GDFR power (P_G). P_G and P_U were measured after removing ASE by using suitable spatial filters. It is seen that in GDFR, as the magnification is increased from 38 to 79, by reducing the diameter of the diffracting aperture from 1.0 mm to 0.7 mm, the percentage of P_G in the total output beam increases reaching a maximum of about 55%. This is due to the fact that as the aperture diameter is reduced in this range, the beam increasingly becomes coherent due to removal of ASE and thus the coherent part extracts more power from the gain medium. For M_G>79 i.e. for 2a < 0.7 mm, the feedback of high quality beam reduces and hence P_G reduces as compared the ASE. The beam however remains coherent across the entire cross-section. In contrast, in case of UR, even though P_U and its percentage in the total output power is higher upto M_U=100, but the coherence width is much smaller than for equivalent GDFR magnification. At M_u=333 the coherence properties nearly match that of GDFR of M_G =38. Comparing the performance of GDFR and UR CVL, it is clear that while the GDFR beam does not show 1/e fall even at low magnification, UR CVL output on the other hand shows markedly low coherence width for equivalent magnifications even though output power is considerably higher than that for GDFR CVL. In UR CVL for M_U > 333 the visibility does not reduce to 1/e value [17].

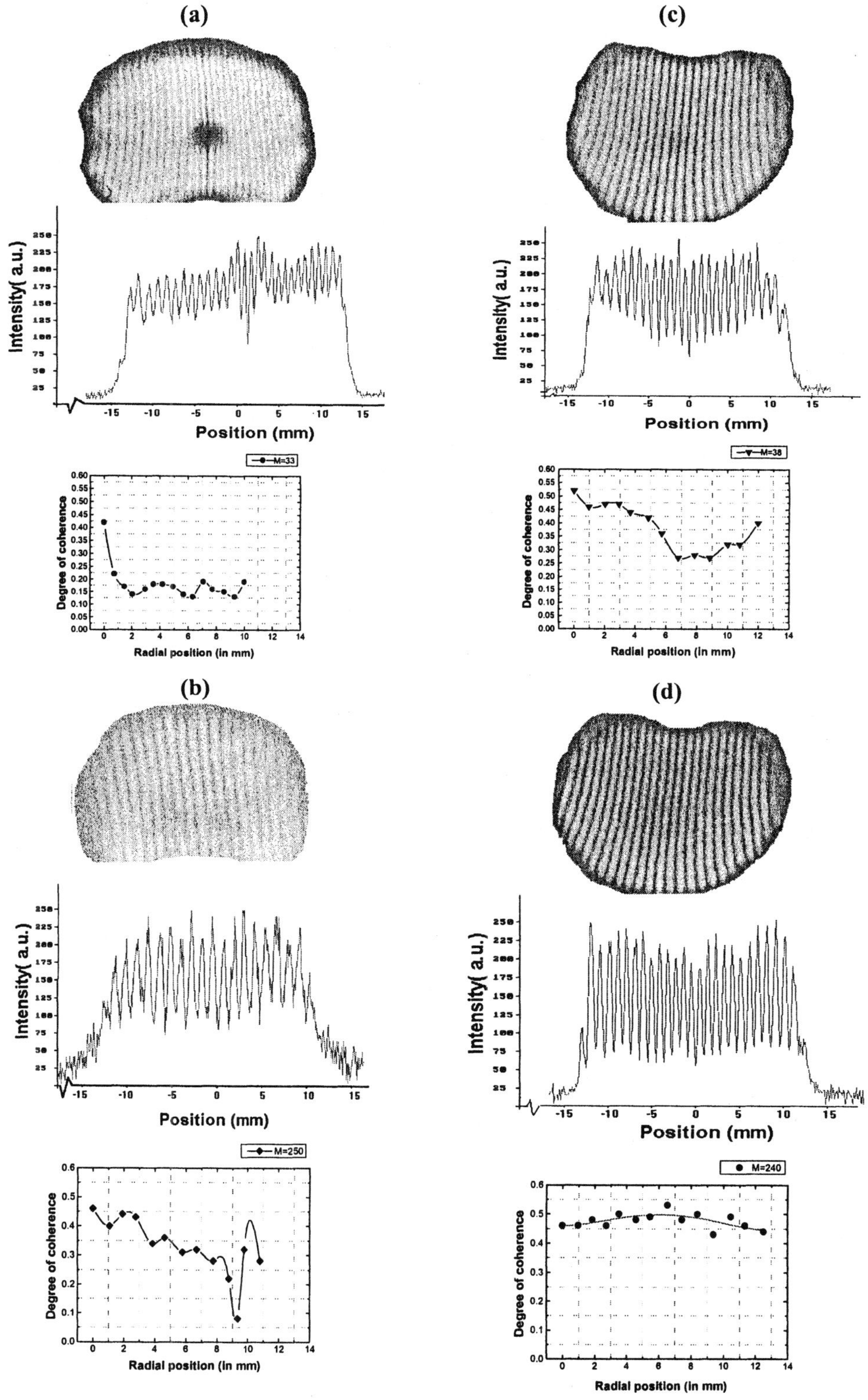

Figure 2: Interferogram, intensity scan and degree of coherence vs radial position for (a) $M_U=33$, (b) $M_U=250$, (c) $M_G=38$ and (d) $M_G=240$.

(b). Time resolved divergence studies of copper vapour laser

The experimental arrangement for time resolved divergence studies is shown in figure (1c). The divergence was estimated by recording the far-field intensity distribution at the focal plane of a lens, L_1, in case of collimated beam from UR or at minimum spot plane in case of GDFR where the beam has additional geometrical divergence (Fig 1c). The magnified image (8 to 14 times) of far-field was formed with a lens L_2 at the slit of streak camera such that it should not exceed the 5mm, size of the photocathode. In all cases, the farfield spot was magnified in such a way that the radiation has propagated about 14 to 15 meters before forming the image at the slit to take care of inherent delay (30 ns) between the trigger pulse and streak image to appear on the screen. The streak camera was operated in the single shot mode. A part of CVL beam was allowed to fall on photodiode which provides the trigger pulse for the camera. In single shot mode operation, streak image will record only 37.5 ns duration. As the pulse duration is longer than single scan of the streak camera an electronic passive delay generator was introduced between the trigger pulse obtained from the photodiode and the trigger input to record the whole pulse duration in two streaks.

Typical streak scan of far field intensity distribution with time are shown in figure (3 a-d) for P-P, UR of M=12.5, 100 and GDFR respectively. The diameter of the beam falling on the lens, L_1 was 28 mm for UR and 33 mm for GDFR. The vertical scale, indicating the farfield spot size, is shown in terms of corresponding diffraction limit, DL (DL=2.44 λ/D = 0.044 mrad for UR and =0.038 mrad for GDFR). From the streak scan it is seen that the farfield spot size, in GDFR, remains constant throughout the pulse while in P-P and UR it falls within the pulse. The divergence was estimated from these streak scans by measuring the farfield spot size ($1/e^2$ point) and dividing by its distance from the lens (Fig. 4). In P-P resonator, the divergence reduced from initial 68 DL to about 30 DL towards the end of the pulse. The divergence reduces by the aspect ratios (i.e. D/2L) in each round trip in P-P resonator.

In case of UR, the divergence at the start of the pulse was 23 DL, 9.8 DL, 4.7 DL and 2.4 DL, for UR M=12.5, 33, 100 and 250 respectively (Fig. 4). There is rapid fall in divergence of UR from 18 ns to 28 ns corresponding to 2nd and 3rd round trip. The magnitude of this fall reduces with increasing magnification. The divergence ultimately reaches 2.7 DL, 2.6 DL, 2.3 DL and 1.7 DL for UR M=12.5, 33, 100 and 250 respectively. It is clear from figure (3) that even for M=250, the divergence does not reach the diffraction limit (DL) even in the last round trip. The calculated divergence of first round trip i.e. the radiation starting from the scrapper mirror side of the gain medium traveling towards the concave mirror and reflected back, is 11.8 mrad (267xDL) for cavity length of 230 cm and focal length of M_1= 250 cm [18]. The divergence of the 1st round trip was very high and its intensity was below the detection limit of the camera and hence could not be detected. In UR CVL, divergence should reduce by a factor of M in each transit. This is not observed. The divergence reduction does not follow the trend except for initial transits in all UR as also observed by Chang et al [14]. At M=250 the divergence reaches only 1.7 DL towards the end of the pulse. The divergence of GDFR

CVL is 1.1 DL at the start and remains constant throughout the pulse (Fig. 4). This is very close to diffraction limit, as predicted theoretically [18].

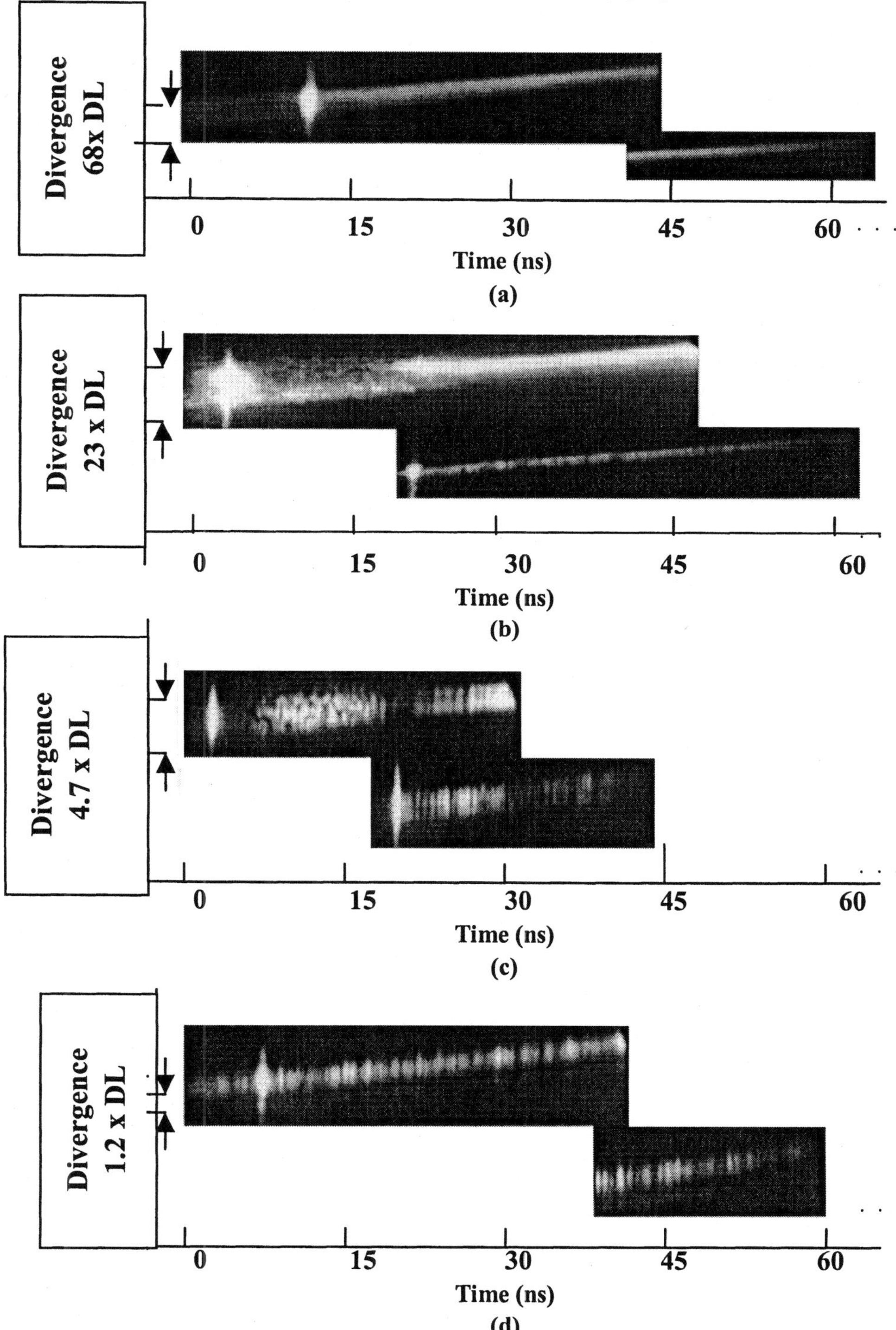

Figure 3: Streak scan of CVL farfield for (a) P-P, (b) UR M=12.5, (c) UR M=100 and (d) GDFR M=107

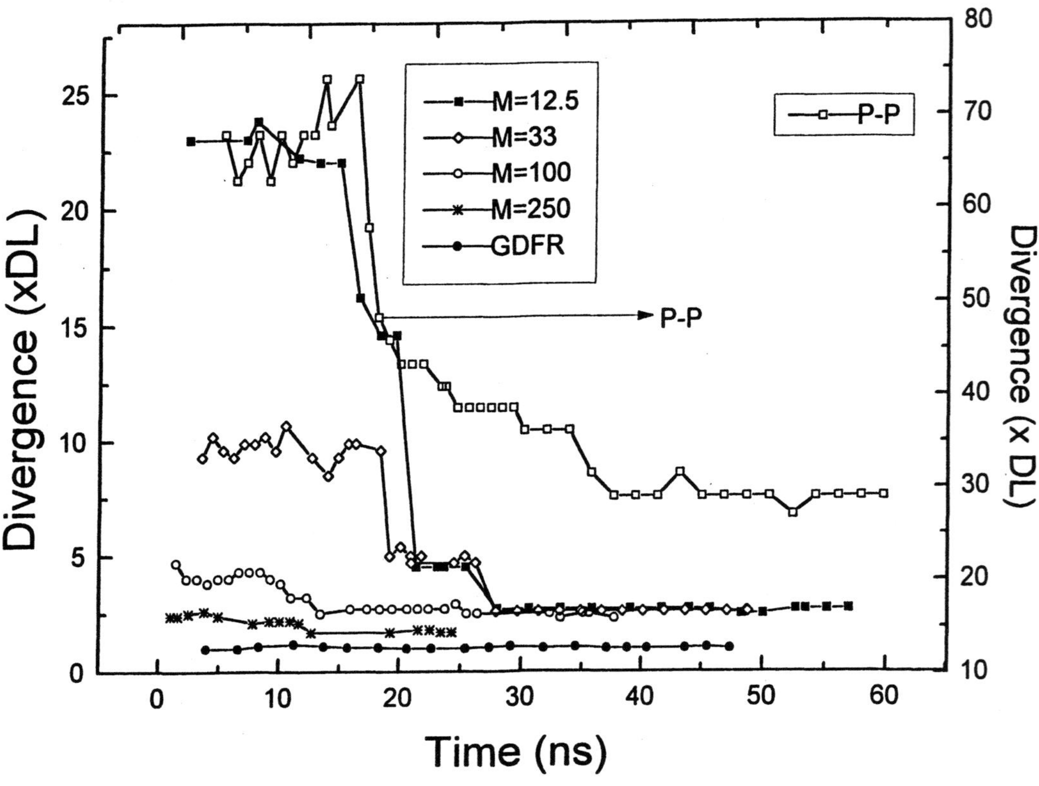

Figure 4: Temporal evolution of divergence within a pulse for various resonators

2   Role of pump beam characteristics in applications

It is seen in the previous section that the difference in the temporal evolution of the beam for different resonators is significant both in terms of behavior and in magnitude. Thus these characteristics are expected to influence the outcome of the applications. Though CVL has been used in many applications, two applications namely second harmonic generation and pumping of a transverse dye laser are taken as representative ones for the study, the results of which are presented below.

(a). Second harmonic generation of CVL beam

The second harmonic generation efficiency depends critically on the spatial coherence of fundamental beam. We have seen in the previous section that GDFR CVL has high degree of spatial coherence which remains constant throughout the pulse. Second harmonic generation of CVL beam using ADP [19], KDP [20]. BBO [1,21] is reported in literature. Using BBO, average conversion efficiencies of 9% to 34% at fundamental powers of 2.5 W to 5.1 W with unstable resonators of magnification 26 to 100 are reported [1, 20-21]. It is expected that with further increase in magnification spatial coherence of CVL will reach diffraction limit, thus promising even increasing efficiencies. We have carried out a comparative study on SHG of CVL with different spatio-temporal characteristics generated by UR of M varying from 12.5 to 250 and GDFR. The collimated beam was line focused by a cylindrical lens in a BBO crystal (M/s Casix, 4 x 4x 7 mm^3) cut at 50° corresponding to type-I phase matching at $\lambda = 510.6$ nm. The crystal was mounted on a 5-axis micropositioner. The fundamental and second harmonic beams were recollimated by a fused silica cylindrical lens and separated by quartz prism.

Figure (5) shows the variation of average SH power as a function of fundamental power for various resonators. Cylindrical lenses of focal lengths 6 cm and 5 cm were found to be optimum for UR and GDFR beams respectively. The SH power increases with increase in UR magnification. The slope also increases with magnification but it was highest for GDFR beam. With GDFR beam 630 mW UV power was obtained at 2.1 W input power (10 % loss of dispersion prism included), whereas for unstable resonator (M=100), input power of 3.5 W was required for the same UV power generation indicating vast difference in efficiency. At M=250, 270 mW UV at input of 1.7 W was obtained. At lower M= 12.5 and 33 the maximum UV output were 290 mW and 480 mW at the input of 4.0 W and 4.5 W respectively.

The conversion efficiencies given as ratio of averaged UV output power and fundamental input power is shown in figure (6). With UR, about the same input power (1.8 W), the highest conversion efficiency was only 15 % (for M=250). Highest overall efficiency was 18.6 % at input of 3.5 W (for M=100). Efficiency of 30 % was observed for GDFR beam at 2.1 W input power. Hence for GDFR, the conversion efficiency at the same input power as well as the highest efficiency is much larger as compared to all the URs studied. The slope of conversion efficiency with fundamental power is highest for the GDFR. In UR, the slope of conversion efficiency increases as the magnification increases. It should be noted that for the same input power, the UV output power increased as UR magnification is increased. As the increase in magnification is related to increase in spatial coherence of the beam, the efficiency of conversion increases with increase in coherence width. With GDFR beam being coherent across its entire crossection (Table), the output power was highest for GDFR beam at same input power.

Figure 7(a-c) shows typical pulse shapes of fundamental, depleted fundamental and UV for UR (M=12.5, 100) and GDFR respectively. It is seen that the depletion increases as the UR magnification increases. Also SH pulse appears earlier with respect to start of the fundamental pulse. In case of GDFR, all the pulse shapes show striking resemblance indicating efficient conversion over almost the entire pulse duration.

Figure 8(a-c) shows the instantaneous conversion efficiency, η $(=P_{2w}/P_w)$ and conversion coefficient C $(C=\eta/P_w$, P_w and P_{2w} being the fundamental peak power and SH peak power respectively) vs time for UR M=12.5, 100 and GDFR respectively. It is seen that in UR, the maximum conversion occurs in the second peak despite the first peak of the fundamental pulse being higher in intensity than the second peak. This is due to larger flux (power/area) for 2nd peak as compared to 1st peak due to large difference in divergence. Optimisation of frequency conversion of UR CVL output poses severe challenge due to temporal evolution of divergence within a pulse. The CVL output of varying intrapulse divergence will be focused in different spot sizes and locations along the axis of the lens. Since the second harmonic generation depends on the fundamental flux incident on the crystal and phase matching conditions, in UR optimum focusing conditions can not be obtained throughout the pulse due to temporally varying divergence. However for GDFR, where divergence was constant throughout the pulse, the flux variation exactly follows the intensity variation. This has resulted in highest conversion efficiencies in the first peak itself. The maximum instantaneous conversion efficiency of 48.6 % was obtained for GDFR in contrast to only 28.6 % from UR M=100. Likewise for UR the conversion coefficient varies from low value to high value within a pulse due to decrease in the beam divergence. The overall conversion coefficient increases as the magnification increases. For GDFR, the conversion coefficient is almost constant throughout the pulse. This is attributed to beam divergence remaining constant within the pulse, which results in optimum focussing geometry being maintained throughout the pulse. In contrast, no single focussing geometry will be optimum throughout the pulse in case of UR.

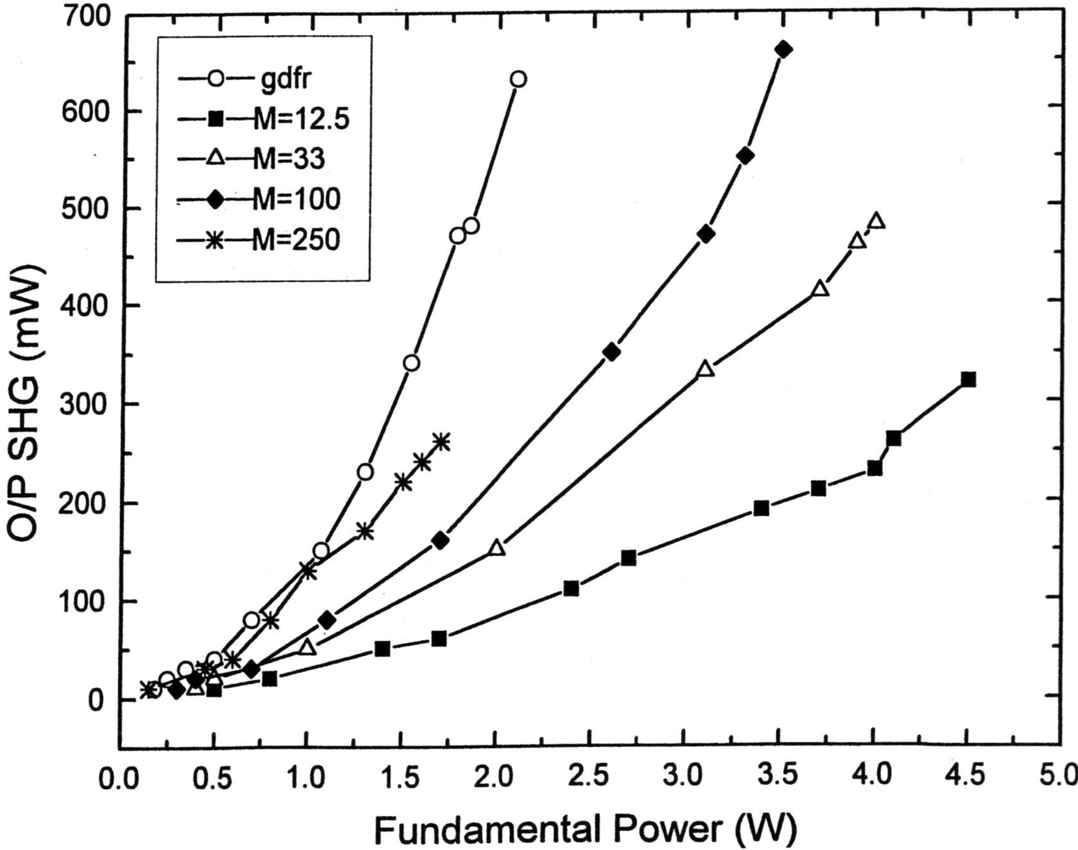

Figure 5: Second harmonic power vs fundamental power for various resonators

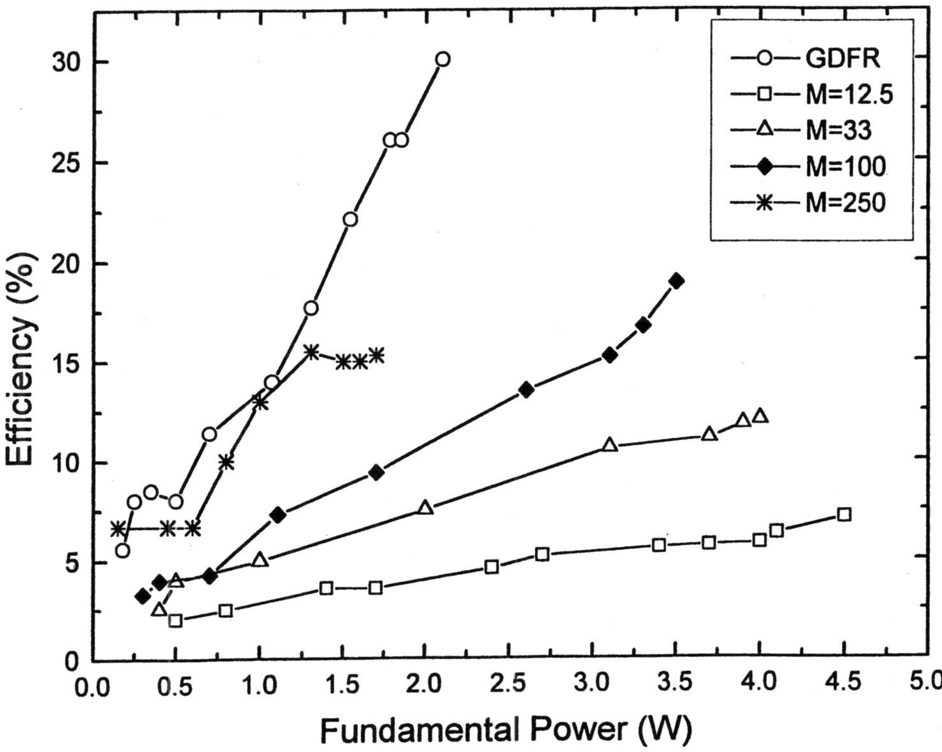

Figure 6: Conversion efficiency vs fundamental input power for various resonators

(b) Transverse pumping of a GIG dye laser

The second application under study is transverse pumping of GIG dye laser where pump beam is not expected to play any role other than producing the active medium. At best, the cross-section of the active region is expected to control the modes of the dye laser. The effect of the transverse mode structure of the pump beam in axially pumped Rh6G dye laser was demonstrated by Corless et al [22]. The dye laser transverse modes follow the transverse modes of the pump beam. In a recent experiment, Ring et al [23] while reporting the results of their experiments on frequency conversion of transversely pumped Rh6G dye laser have observed that conversion efficiency is higher when dye laser pumped by an unstable resonator CVL beam as compared to the case when the resonator was plane -plane. However no explanation was offered. This could only be possible if the pump beam was affecting the divergence and bandwidth of the dye laser. There are two way in which the coherence of pump beam could effect the dye laser divergence and hence the bandwidth, namely through control of modes due to gain medium acting as an aperture [24] and secondly through coherent excitation of dye molecules which may give rise to increased coherence of the seed radiation instead of

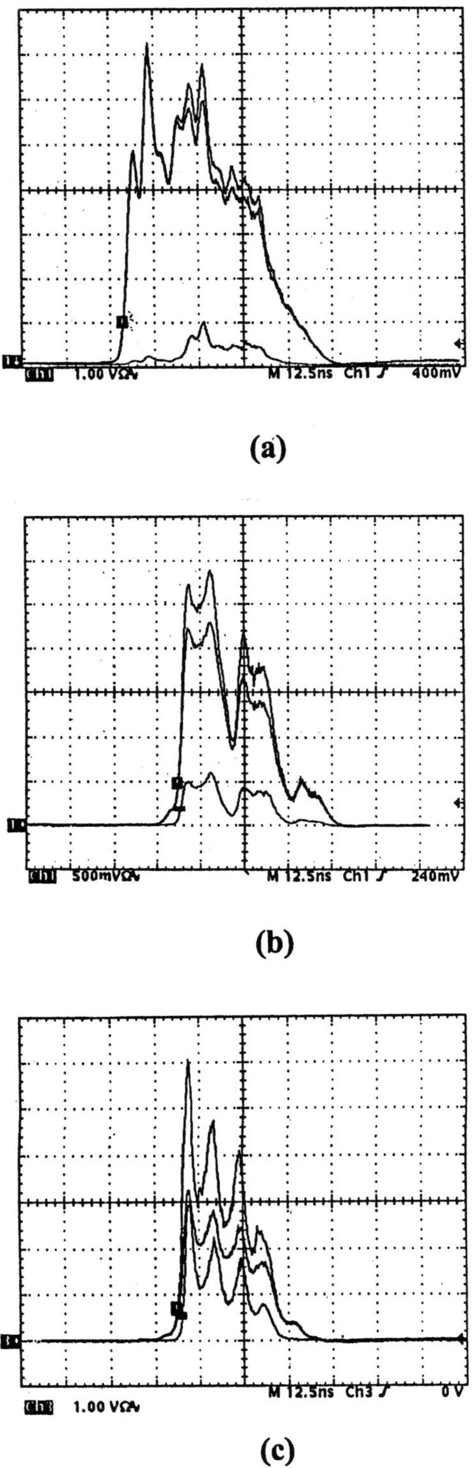

(a)

(b)

(c)

Figure 7: Fundamental (upper trace), depleted fundamental (middle trace) and SHG (lower trace) pulse shapes for   (a) UR M=12.5 (b) UR M=100 (c) GDFR

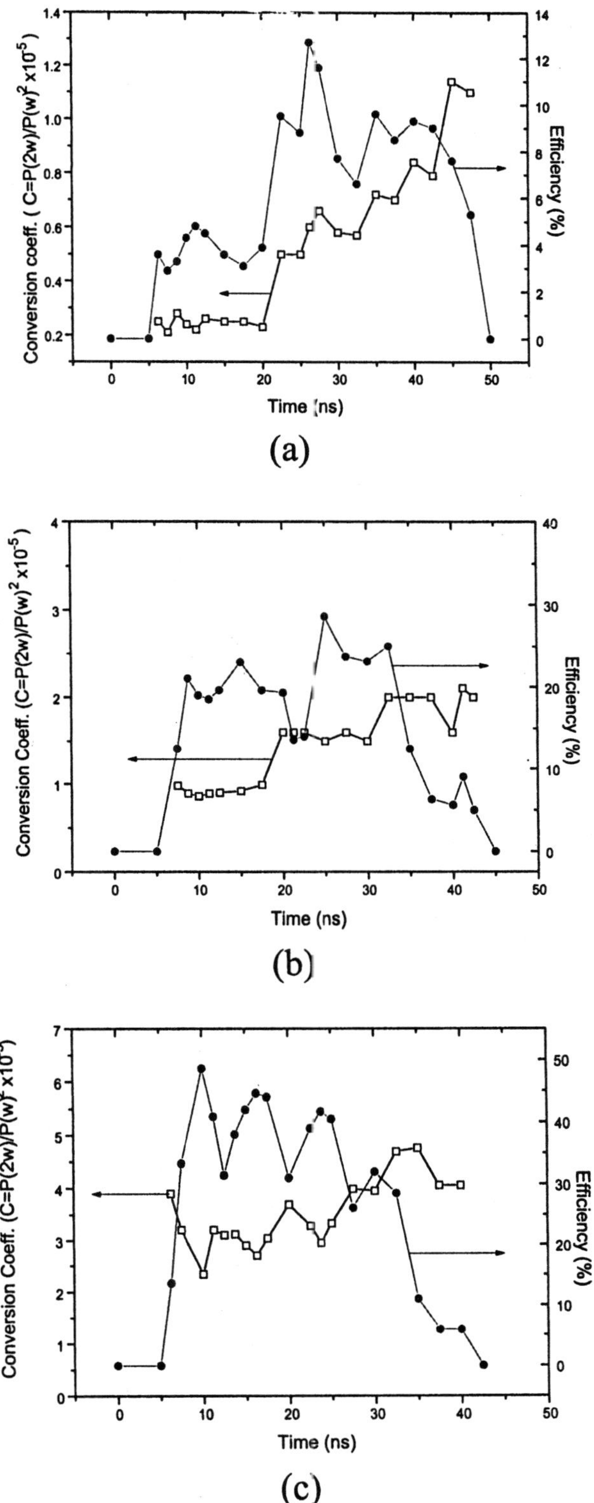

Figure 8: Conversion efficiency and coefficient vs time for (a) UR M=12.5 (b) UR M=100 (c) GDFR

buildup from an incoherent seed radiation. It is known that under excitation by coherent field the excited atoms/molecules emit coherently [25]. It is therefore possible that the coherence of the seed emission responsible for dye laser build up may eventually control its characteristics. As seen in the previous section, wide difference in the characteristics of output beams from different resonators offer good opportunity of exploring this aspect in transverse pumped dye laser characteristics. Authors in a recent paper have shown that the pump beam coherence/divergence indeed effects the dye laser divergence and bandwidth [26]. Some of the results of the findings are presented here.

Of the several resonators configurations discussed in the previous section three representative resonators were selected. Of these two resonators namely plane-plane and low magnification UR are used widely for dye laser pumping [27-28]. These have poor to moderate coherence/ divergence characteristics. The third, namely GDFR [5] has so far demonstrated highest coherence and was used for dye laser pumping for the first time [26]. The pump beams are designated as R_P, R_U and R_G for P-P, UR M=12.5 and GDFR respectively. The dye laser chosen is a two -prism beam expander (magnification=5.6), grazing incidence grating (80^0, 2400 lines/mm), transversely pumped (f= 7 cm) Rh6G laser (25 mm gain length, 2.0 mM in ethanol) [29]. The development of the dye laser characteristics within a single pulse are studied using a nanosecond resolution streak camera with same time resolution as used in CVL pulse studies. All the experiments were carried out at same wavelength which was monitored with a 0.5m monochromator to ensure that the grating angle did not change. In fact except the focusing lens no optical component was disturbed.

Figure 9(a-c) shows the temporal variation of F-P (FSR=20 GHz, Finesse=14) fringe pattern for R_P, R_U and R_G pump. The width and separation between the fringes is carefully monitored. The start of CVL pulse is taken as zero time reference. In case of R_P pump, an additional fringe starts appearing and the fringe spacing also changes indicating change in the bandwidth (Fig. 9a). In case of R_G pump, the F-P ring separation is almost constant throughout the pulse (Fig. 9c). Variation in bandwidth is plotted in figure (10) for all the pump beams. It is observed that the bandwidth for R_U pump lies between extreme conditions of low coherence width, R_P pump and high coherence width, R_G pump. The bandwidth and its variation decreases as the pump laser is changed from R_P to R_G. Also the percentage change during the pulse is minimum in case R_G, where the divergence is constant. It is also seen that the bandwidth reduces for each case within the dye laser pulse.

In order to understand this further, we studied temporal variation of the dye laser divergence during the single pulse. Far-field intensity distributions were recorded at the focal plane of a lens of 20 cm focal length for R_P -R_G pump and are shown in figure (11.a-c) respectively. The divergence, estimated from these, is plotted in figure (12). It is seen that the variation is more in case of the pump beam with higher divergence (R_P) than that for R_G beam. The divergence for R_U pump beam lies between that for R_P and R_G pump beams. The divergence for R_P and R_U pumping decreases to a large extent towards the end of the pulse as compared to R_G which shows slight decrease in divergence. The trends are reflected in the bandwidth measurements (Fig. 10).

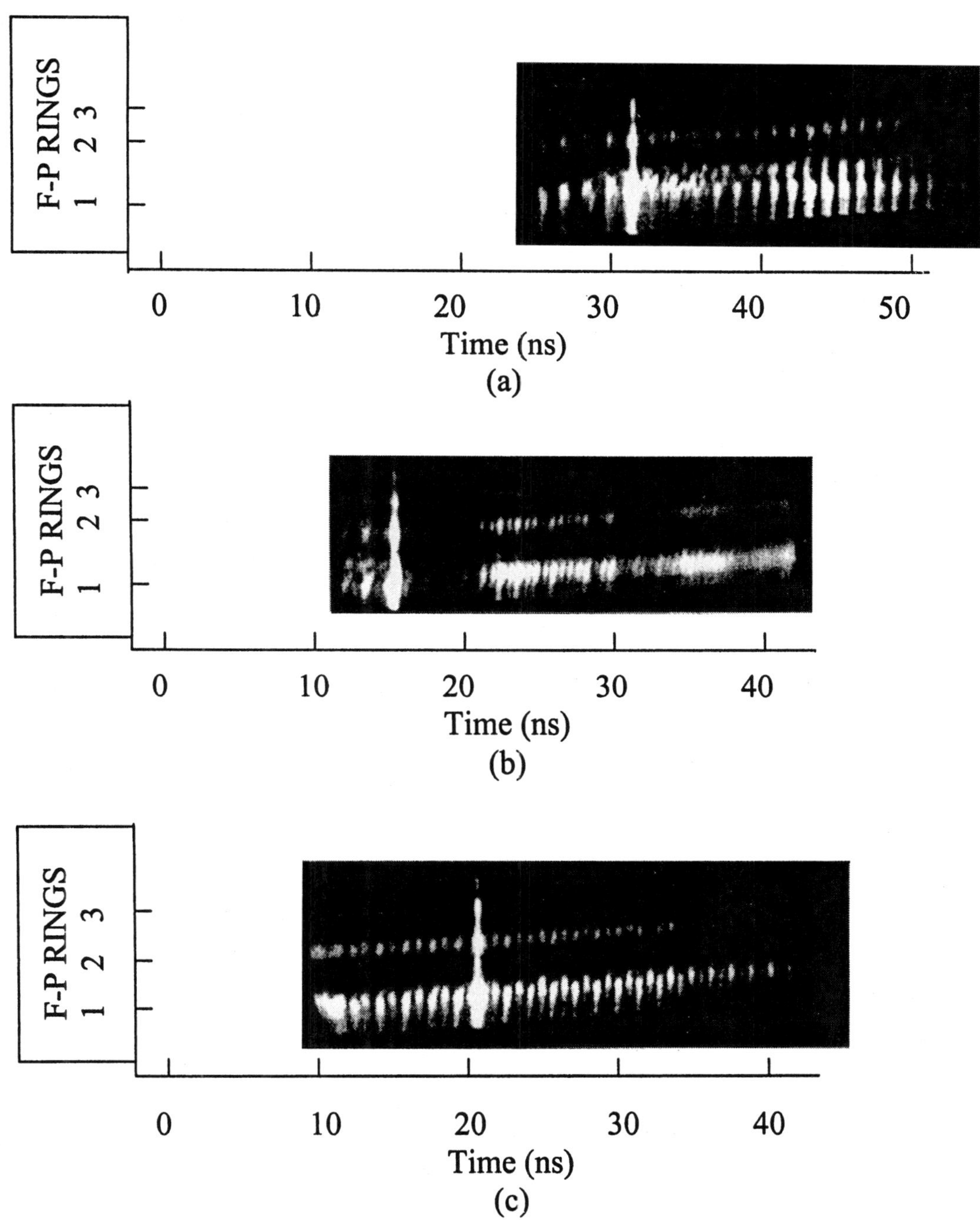

Figure 9: Single pulse streak scan of F-P ring within a dye laser pulse for pump beam (a) PP (b) UR (c) GDFR

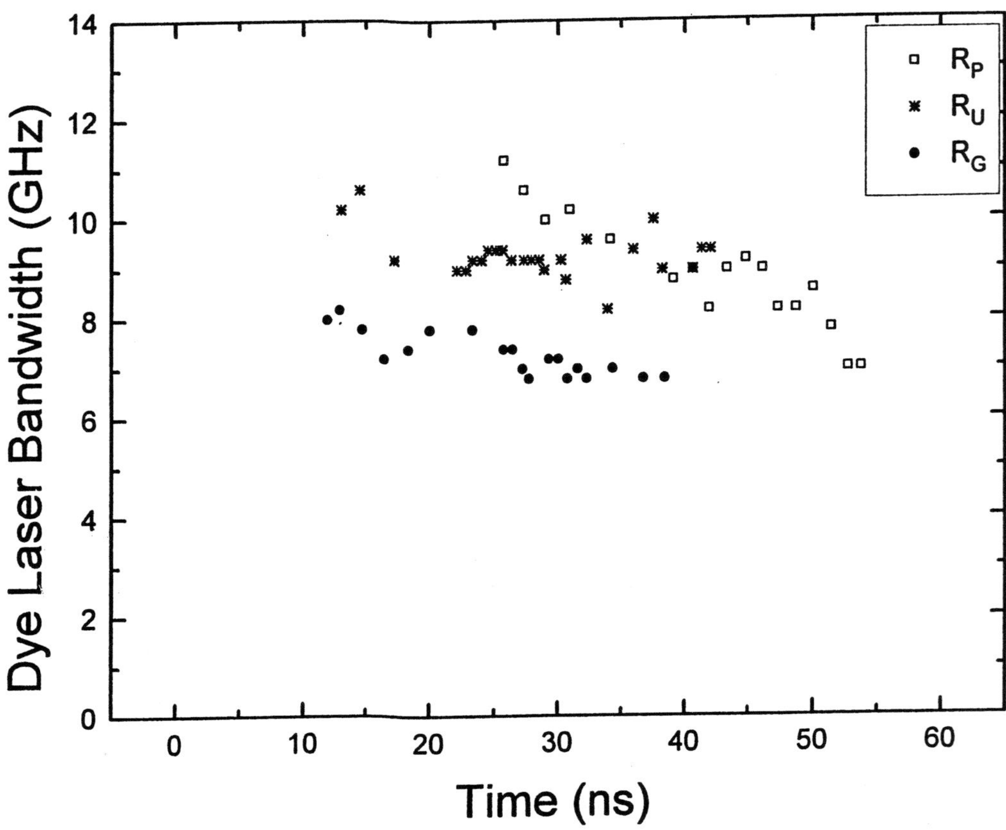

Figure 10: Variation of dye laser bandwidth vs time for different pump beams P-P (R_P), UR (R_U) and GDFR (R_G)

The divergence, $\Delta\theta$, depends on the source size, which is nothing but the size of the gain medium, i.e. the penetration depth of pump beam into dye medium [30]. The variation of dye medium size with time for R_P, R_U, and R_G pumped dye laser is shown in figure 13(a-c) respectively. The dye medium size was estimated as the depth at which intensity falls to $1/e^2$ of the peak intensity (Fig. 14). The gain medium size varies from 214 µm to 123 µm in R_P, 220 µm to 142 µm in R_U, and 240 µm to 170 µm in R_G. This size is incidentally larger by a factor of 3-4 to the value estimated from Bear-Lambert Law [30]. It is seen that the gain medium size also varies within the pulse decreasing toward the end. Decrease in the gain medium size will result in smaller beam waist leading to oscillations in lower order modes and hence decrease in divergence $\{\Delta\theta(t) = \lambda / \pi\omega, \ \omega$ being the waist size$\}$. The dye laser divergence, (Fig. 12) and the bandwidth, (Fig. 10) indicate direct relation with the pump beam divergence (Fig 4). Lower the divergence of the pump beam, lower is the dye laser divergence and bandwidth.

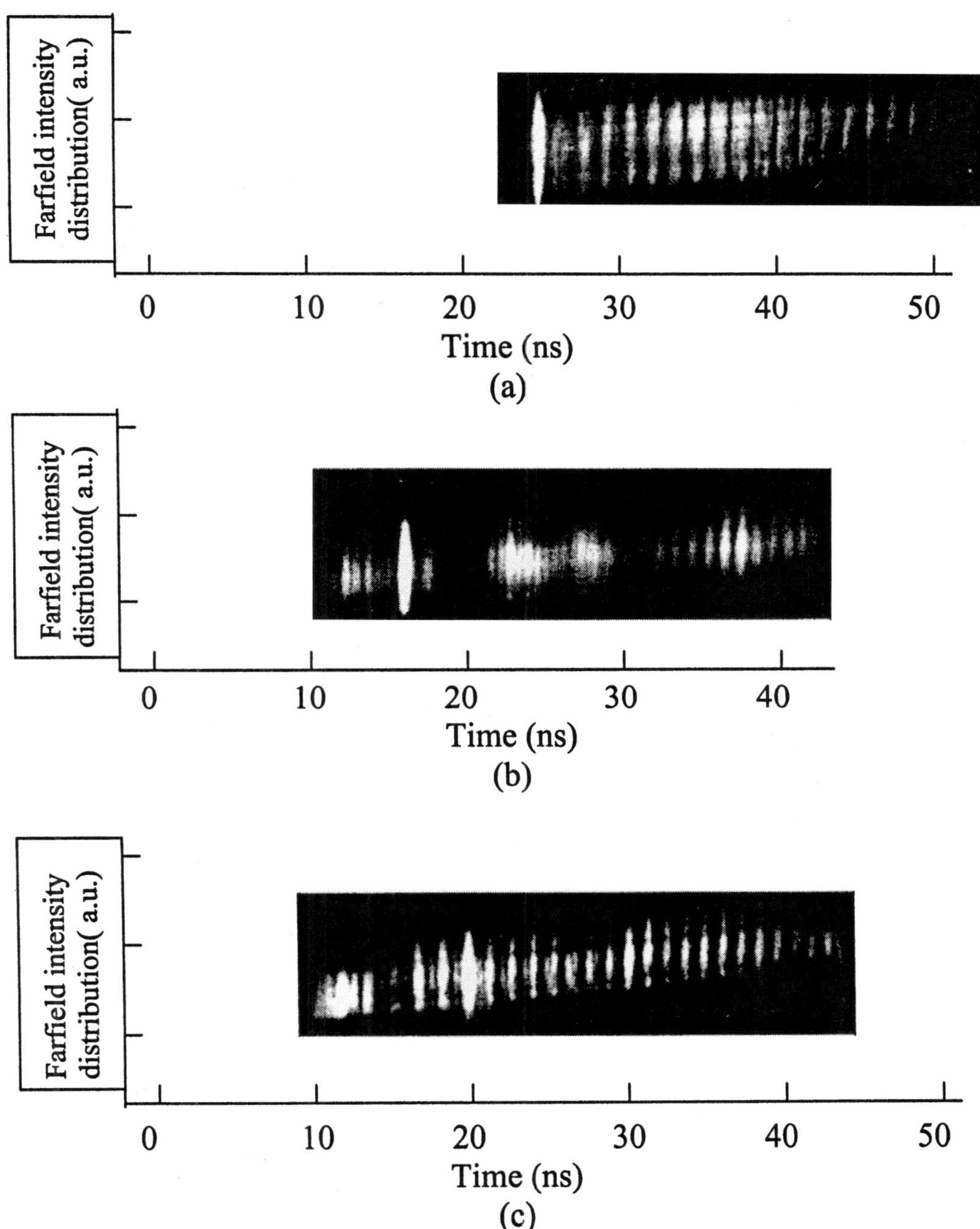

Figure 11: Single pulse streak scan of dye laser far field spot for pump beam with resonators (a) PP (b) UR (c) GDFR

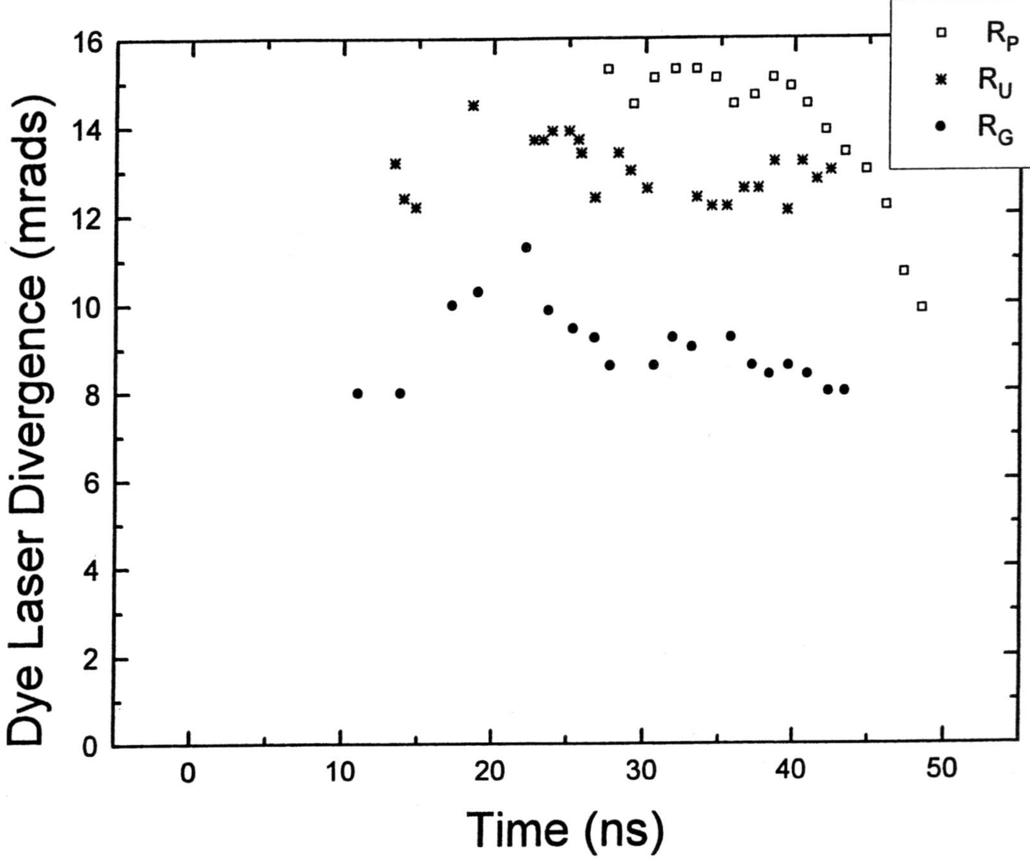

Figure 12: Variation of dye laser divergence vs time for different pump beams
P-P (R_P), UR (R_U) and GDFR (R_G)

It is observed from figure (13) that in the time interval from 10 to about 30 nsec the dye medium size is same for all the three cases. Thus during this time the divergence (Fig.12) and the bandwidth (Fig.10) of the dye laser should be the same for all resonators if it is dominated by the medium aperture alone. However this is not the case. Divergence and the bandwidth both are minimum for GDFR beam and largest for P-P with values for UR lying in between. Referring to Table again, it can be seen that with coherence width for GDFR is about 28 mm. When focussed by a cylindrical lens, the dye molecules in full length (25 mm) of the dye medium are under excitation by a beam which is temporally and spatially coherent. The dye radiation therefore builds up from higher coherence in the emission. As compared to this, UR (R_U) beam has coherence width of only 3.6 mm in a beam of about 28 mm crossection leading to only a small number of dye molecules under the effect of coherent excitation. Since the coherence width for plane-plane resonators R_P was not measurable the dye emission starts with poor coherence. Thus high tempo-spatial coherence in case of GDFR (R_G) pump beam results in build up of coherent emission from the dye, resulting in lower divergence and hence lower bandwidth as compared to pumping by beams from other two resonators (R_U & R_P) even though the dye medium size is the same. Decrease in bandwidth towards the end coupled with decreased in spot size indicate mode control by the dye medium.

Conclusions

In conclusion, it is seen that the in UR the coherence width increases as the magnification increases. For CVL, GDFR offers better spatial coherence at all magnifications as compared to UR. The GDFR beam is coherent across its diameter even at low magnifications and output power is twice as high as that for UR CVL for a fully coherent beam. The variation of divergence within a pulse decreases as the magnification is increased whereas no variation was observed for the GDFR. Effect of pump beam characteristics on two of the commonly used applications namely second harmonic generation and transverse pumping of a dye laser was studied and results reported. It is shown that the instantaneous conversion efficiency was higher throughout the pulse in case of GDFR as compared to that for UR. This was attributed to optimum focusing conditions being maintained in case of GDFR rather than in case of UR.

In case of dye laser pumping, it is shown that the bandwidth of the dye laser is controlled during the rising and falling part of the pulse, by the size of the gain medium which changes with change in flux. For the same size of the dye medium, the bandwidth depends on the coherence width of the pump pulse. Higher is the coherence width lower is the bandwidth. Variation of bandwidth and divergence of the dye laser within the pulse are least for GDFR compared other resonators.

Acknowledgement

The authors wish to acknowledgement the assistance and cooperation of all CVL group members who were responsible for the development of copper vapour laser and dye laser.

References

[1]. K. Kuroda, T. Omastu, T. Shimura, M. Chihara and I. Ogura, Opt. Comm., 75 (1990) 42.

[2]. R. Kufer and H. Bergman, Proc. Soc. Photo-opt. Instrum. Eng., 1276 (1991) 411.

[3]. R. S. Hargove and T. Kan, IEEE J. Quantum Electron., QE-16 (1980) 1108.

[4]. R. Bhatnagar, S. K. Dixit, P. K. Shukla, B. Singh and J. K. Mittal, in Pulsed Metal Vapour Laser, C. E Little and N. V. Sabotinov eds., (Kluwer Academic Publishers, Netherlands, 1996), page 255-262.

[5]. S. K. Dixit, J. K. Mittal, B. Singh, P. Saxena and R. Bhatnagar, Opt. Comm., 98 (1993) 91.

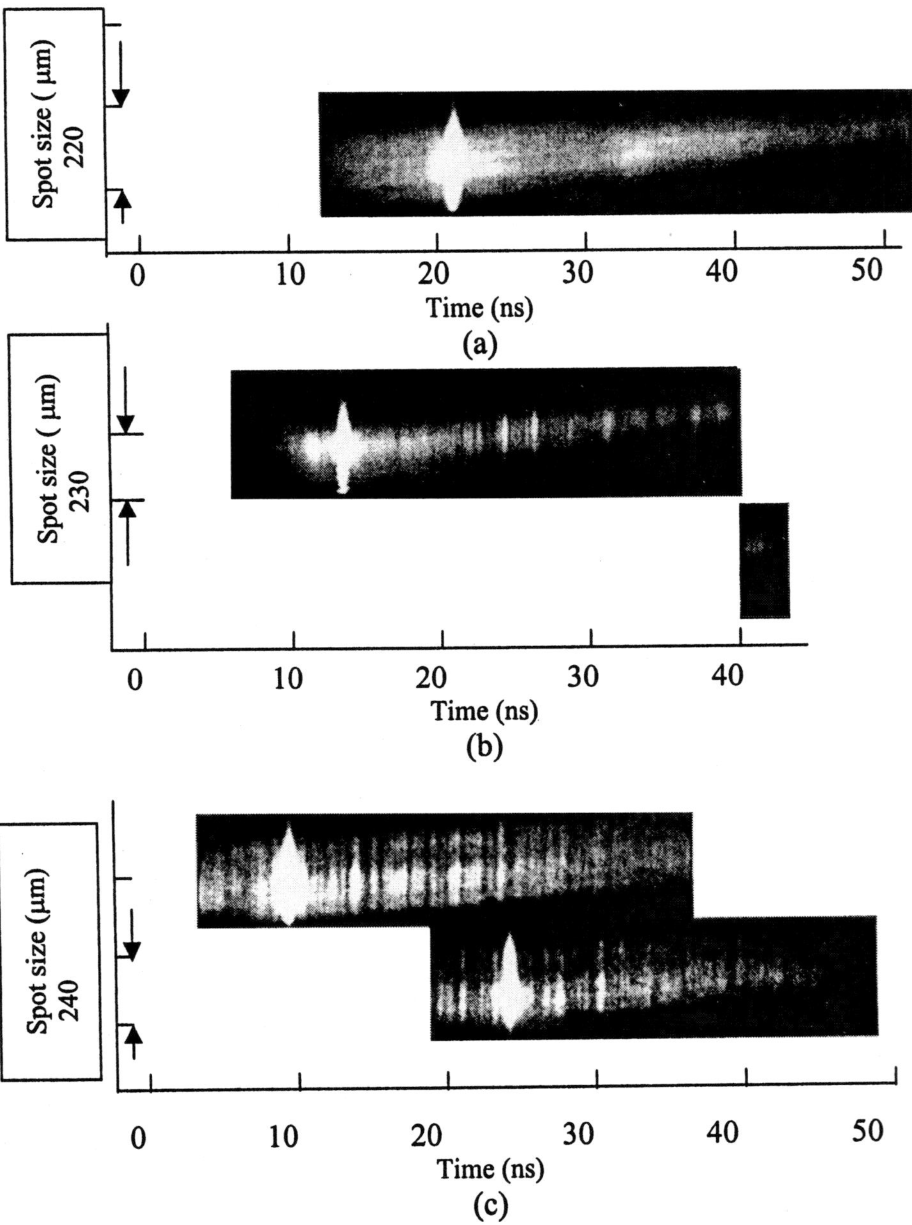

Figure 13: Streak scan of dye medium size within single CVL pulse for pump beam with resonator (a) PP (b) UR (c) GDFR

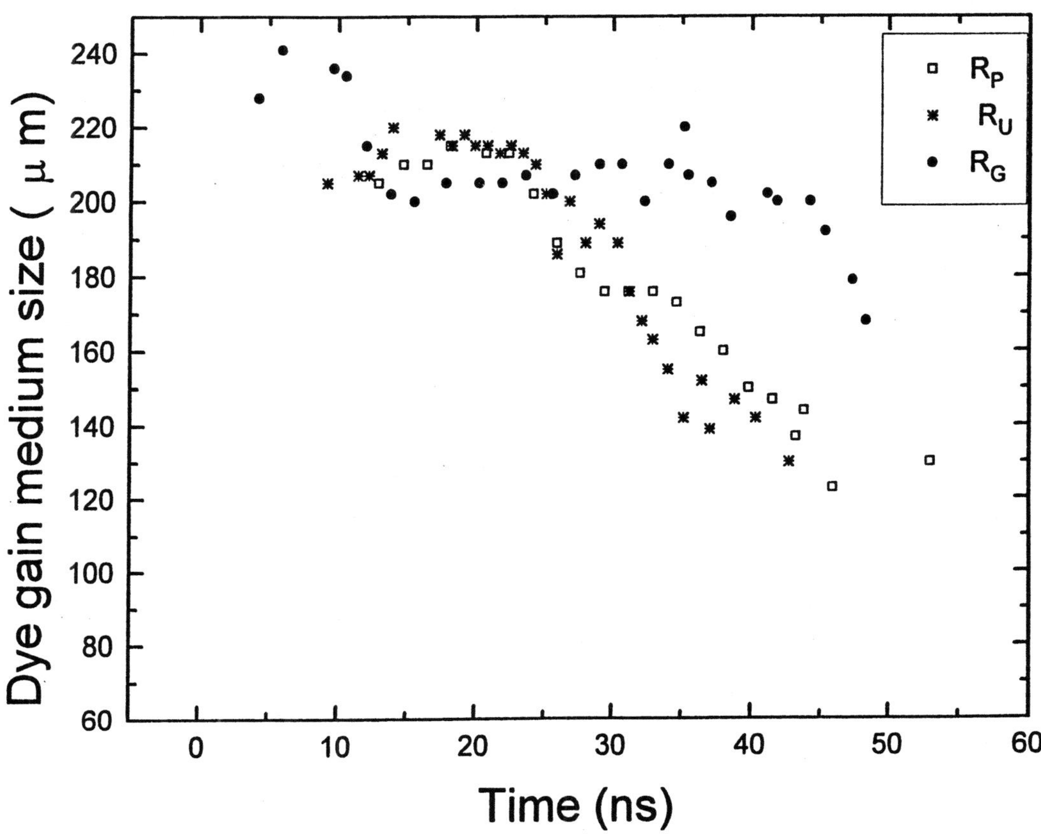

Figure 14: Variation of dye gain medium size vs time for different pump beams
P-P (R_P), UR (R_U) and GDFR (R_G)

[6]. R. S. Hargrove, R. Grove and T. Kan, IEEE J. Quantum Electron., 15 (1979) 1228.
[7]. V. P. Belyaev, V. V. Zubove, A. A. Isaev, N. A. Lyabin, Yu. F. Sobolev and A. D. Chursin, Sov. J. Quantum. Electron., 15 (1985) 40.
[8]. V. V. Zubove, N. A. Lyabin and A. D. Chursin, Sov. J. Quantum Electron., 16 (1986) 1606.
[9]. M. Amit, S. Lavi, G. Erez and E. Miron, Opt. Comm., 62 (1987) 110.
[10]. T. Omatsu, K. Kuroda and T. Takase, Opt. Comm., 87 (1992) 278.
[11]. D. W. Coutts, M. D. Ainsworth and J. A. Piper, Opt. Comm., 87 (1992) 245.
[12]. D. W. Coutts, IEEE J. Qunatum Electron., 31 (1995) 330.
[13]. T. Omastsu, T. Takase and K. Kuroda, Opt. Comm., 97 (1993) 65.
[14]. J. J. Chang, Appl. Opts, 33 (1994) 2255.
[15]. T. Omastu, K. Kuroda, T. Shimura, K. Chihara, M. Itoh and I. Ogura, Opt. and Quant. Electron., 23 (1991) S447.

[16]. H. S. Vora, S. V. Nakhe, K. K. Sharangpani, P. Saxena, R. Bhatnagar and N. D. Shirke, Proc. National Laser Symposium, U. Nundy, ed. (National laser program, Centre for Advanced Technology, Indore, India), 1994, 260.

[17]. Om Prakash, P. K. Shukia, S. K. Dixit, S. Chatterjee, H. S. Vora and R. Bhatnagar, Apl. Opt, 37 (1998) 7752.

[18]. S. K. Dixit, B. Singh, J. K. Mittal, R. Choube and R. Bhatnagar, Opt. Eng., 33 (1994) 1908.

[19]. A. A. Isaev, G. Yu. Lemmerman and G. L. Malafeeva, Sov. J. Quantum Electron., 10 (1980) 983.

[20]. Yu. P. Polunin and V. O. Troitskii, Sov. J. Quantum Electron., 17 (1987) 1433.

[21]. D. W. Coutts and D. J. W. Brown, IEEE J. Sel. Top. Quantum. Electron., 1 (1995) 768.

[22]. J. D. Corless, J. A. West, J. Bromage and C. R. Stroud, Jr. Rev. Sci. Instrum, 68 (1997) 2259.

[23]. H. Ring, U. Rasbatch, L. Michaille, S. Buscher and W. Urban, J. P. Pique, Opt. Comm., 156 (1998) 178.

[24]. S. K. Dixit, S. R. Daulatabad, P. K. Shukla and R. Bhatnagar, Opt. Comm., 134 (1997) 149.

[25]. A. E. Siegman, Lasers, University Science, Mill Valley, California, 1986), pp. 133.

[26]. Om Prakash, S. K. Dixit, R. Jain, S. V. Nakhe and R. Bhatnagar, Opt. Comm, 176 (2000) 177.

[27]. A. A. Pease and W. M. Pearson, Apl. Opt., 16 (1977) 57.

[28]. A. F. Bernhardt and P. Rasmussen, Apl. Phys. B, 26 (1981) 141.

[29]. F. J. Duarte and James A. Piper, Apl. Opt., 23 (1984) 1391.

[30]. F. J. Duarte, in Dye Laser Principle, F. J. Duarte and L. W. Hillman, eds. (Academic press. New York, 1990) pp. 133-183.

Optical Emission Diagnostics of Laser Produced Carbon Plasma

S. S. Harilal*
International School of Photonics
Cochin University of Science & Technology, Cochin 682 002, India
Present Address : Institut Fr Experimentalphysik V
Ruhr Universitt Bochum, D-44780, Bochum, Germany

1 Introduction

With the advent of high power pulsed lasers, the study of the interaction of light with matter has assumed a new dimensions in the past three decades [1]. The subject has made significant progress in its application to many fields of basic research and material technology such as thin film deposition, production of clusters and nanoparticles, lithography, etching, annealing and in fabrication of microelectronic devices [2, 3, 4, 5, 6] The high luminous plasma produced during and after laser - target interaction is a very rich source of ionic, atomic and molecular species and clusters of the atoms. Considerable efforts are being made worldwide in plasma diagnostics and in studying dynamics of laser target interactions. The various phenomena like laser plasma interaction [7, 8], cluster formation and dissociation [9, 10], gas phase chemical reactions [11, 12], plasma ionization [13, 14] etc. make it rather difficult to understand the evolution and dynamics of laser ablation completely. Mechanisms of laser ablation varies widely with the mechanical, thermal and absorptive properties of the material as well as the laser parameters and ambient conditions [15]. Gas phase collisions inside the plasma plays a major role in determining the spatio-temporal properties [16, 17].

When the particle densities in a plasma become so high enough, they undergo collisions in the high density region near the target forming a so called Knudsen layer. Kelly and co-workers extensively investigated the formation of such a Kundsen layer (KL) [18, 19, 20, 21]. The formation of KL results in stopped or backward moving material close to the target and strongly forward peaked velocity distributions away from the target. The KL formation followed a more forward peaked particle flux with an unsteady adiabatic expansion (UAE) of the plasma. Lasers

---

*E-mail : harilal@ep5.ruhr-uni-bochum.de

have been used to produce high temperature and high density plasmas by vaporizing a small amount of material with high powered picosecond or nanosecond pulses [22, 23]. In the intense laser field of picosecond laser pulse, various processes like tunnel ionization and relativistic skin effect etc.. appear in addition to multiphoton ionisation and collision induced ionization whereas in the nanosecond regime the ionization inside the plasma is limited mainly to the latter effects. Laser - solid interaction causes several processes to occur simultaneously and these include energy transfer, evaporation of solid material, generation of a dense plasma, formation of charged and neutral particles. Laser induced plasma has a very short temporal existence and is transient in nature, with a fast evolution of the characteristic parameters that are heavily dependent on irradiation conditions such as incident laser intensity, irradiation spot size, ambient gas composition and pressure [7, 24, 25]. Detailed investigation of the optical emission of the plasma plume gives information on the spatial and temporal evolution of transient species produced during laser-target interaction, such as excited atoms, ions or molecules [26, 27]. Pulsed laser ablation(PLA) is a technique utilized for thin film preparation, in which a laser beam is focussed onto a solid target and the resulting vapour is made to condense onto the substrate of interest. The advantages of PLA are its ability to replicate the stoichiometry of the ablation target within the laser deposited film and the high energy of the ablated species leads to formation of good quality thin films [3, 28, 29, 30].

Usually plasmas are characterized by their electron temperature (T_e) and electron density (n_e). There are several diagnostic techniques employed for the determination of electron density which includes, plasma spectroscopy [31, 32, 33], Langmuir probe [34, 35, 36], microwave and laser interferometry [37, 38], and Thomson scattering [39, 40]. Thomson scattering is probably the most direct and least theory-dependent, while spectroscopy is the simplest as far as instrumentation is concerned [41]. Plasma density determination using Stark broadening of spectral lines is a well established and reliable technique in the range of number density 10^{14} to $10^{18} cm^{-3}$ [42, 43, 44]. The electron temperature is an equally important plasma parameter. Knowledge of the electron temperature is vital to the understanding of the dissociation, atomization, ionization and excitation processes occurring in the plasma. It can be determined spectroscopically in a variety of ways: from the ratio of integrated line intensities, from the ratio of line intensity to underlying continuum, and from the shape of the continuum spectrum (bremsstrahlung) [33].

The interaction of high power laser beam with a graphite target has become a subject of great interest since the first report on the existence of stable carbon clusters called fullerenes in the laser ablated graphite plasma [45]. Carbon clusters like C_{60} and higher fullerenes are well known to be formed as a product of the laser ablation of graphite in an ambient helium atmosphere [46]. Carbon molecules (C_n) are particularly interesting due to their unique and fascinating structural and spectroscopic properties, their importance in astrophysical processes and due to their role in combustion and soot formation. Laser ablation has the unique advantage that most of these molecules are formed in their excited states and hence spectroscopic measurements offer an excellent means to investigate their evolution and dynamics [47, 48]. Another important application of laser induced carbon plasma is in the production of diamond-like carbon (DLC) thin films. DLC films have wide applications owing to their interesting characteristics like extreme hardness, electrical insulation, high thermal conductivity, optical transparency etc. With pulsed laser evaporation of carbon targets hydrogen free films with improved characteristics may be grown [49]. Several types of lasers have been used for the preparation of DLC films which include excimer lasers [50, 51], Nd:YAG laser [52, 53], CO_2 [54] with laser intensities varying from 0.1 to 100 GW cm^{-2}. Recently, a critical review of the pulsed laser deposition of amorphous DLC films is performed by Voevodin and Donley [55]. Creasy and Brenna [56] have reported fullerene ions from laser ablation of DLC films demonstrating that C_{60} ions can be formed in a variety of systems.

Several spectroscopic studies of graphite plasma have been carried out using a variety of laser wavelengths such as 193, 248, 308, 532 and 1064 nm [57]. It has been shown that shorter wavelengths are more effective for penetration into the sample, mainly because of large ablation rates possible at these wavelengths [58]. However, the main advantage in the use of NIR low energy photons is that, they are less likely to invoke photochemistry into the ablation phenomenon. Stevefelt and Collins [59] reported a modelling of the free expansion of a carbon plasma produced at moderate intensities, which includes the dominant constituent mechanisms of electron-ion recombination through both three body collisional and radiative processes. In the context of laser deposition of DLC films, it has been reported that high quality DLC films are obtained at moderate laser irradiances where molecular C_2 formation is prominent as revealed by its emission spectrum [60]. Thus for the production of DLC films it is necessary to understand the formation and dominance of C_2 species and its dependence on various parameters like laser irradiance, pressure of the ambient gas, concentration of the species at different distances normal to the target etc.. Despite considerable experimental and theoretical progress, the studies on laser produced carbon plasma have not yet yielded a clear-cut picture on plasma dynamics of the cluster formation and such a situation arises mainly due to the complexity of the phenomena involved. In this article, the details of a time and space resolved optical emission spectroscopic investigations on laser induced carbon plasma are given. In the first section the variation of two fundamental parameters of the carbon plasma, viz. electron density and temperature with various experimental parameters are discussed. In the preceding section, a special emphasis is given for the time of flight (TOF) features of C_2 emission spectrum under various experimental conditions.

2 Experimental Set up

A schematic diagram of the of the experimental set up is given in figure 1 for spectroscopic studies. The carbon plasma was generated by laser ablation of the

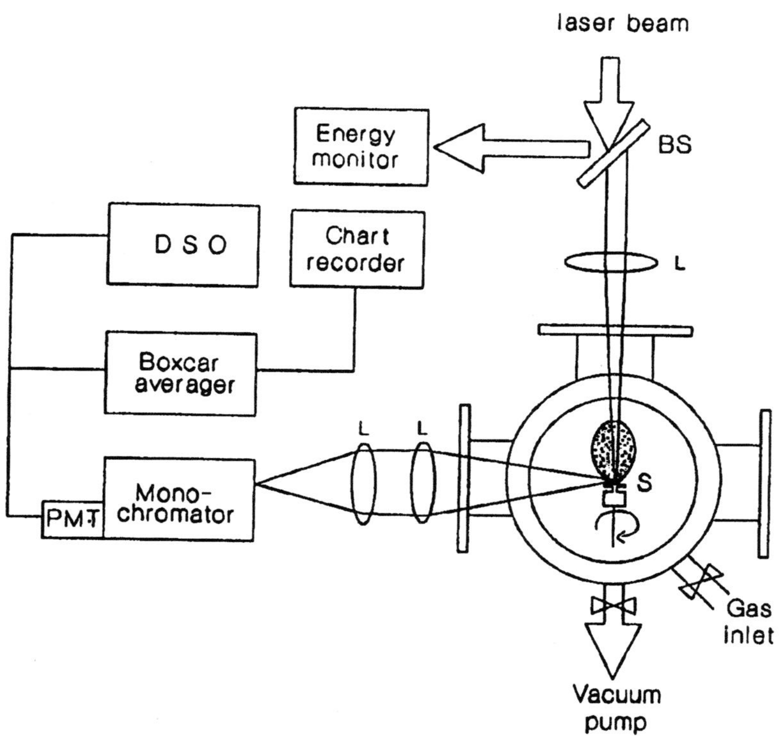

Figure 1: A block diagram of the experimental arrangement for the emission spectroscopy (BS,beam splitter; S, sample; L, lens; DSO, digital storage oscilloscope)

high purity polycrystalline graphite sample using 1064 nm radiation pulses from a Q-switched Nd:YAG laser with repetition rate 10 Hz. The target in the form of a disc (20 mm diameter and 5 mm thickness) was placed in a evacuated chamber provided with optical windows for laser irradiation and spectroscopic observation of the plasma produced from the target. The target was rotated about an axis parallel to the laser beam to avoid nonuniform pitting of the target surface. The bright plasma emission was viewed through a side window at right angles to the plasma expansion direction. The section of the plasma was imaged onto the slit of a 1 metre Spex monochromator (with entrance and exit slits are parallel to the target surface) using appropriate collimating and focusing lenses so as to have one to one correspondence with the sampled area of the plasma and the image. The scan of the monochromator was controlled using Spex CD2A compudrive arrangement. The recording was done by using a thermoelectrically cooled photomultiplier tube (PMT), which was coupled to a boxcar averager/gated integrator. The averaged output from the boxcar averager was fed to a chart recorder, which for the present study averaged out intensities from 10 pulses. For Stark broadening studies, the resolution of the monochromator was kept at its maximum by keeping the slit width at a minimum. For time resolved spectroscopic studies the output from the PMT is directly coupled to a 200MHz digital storage oscilloscope with 50 Ω termination.

3 Electron Density and Temperature Measurements

Among the various plasma parameters electron temperature and density are the main quantities that determine the characteristics of the plasma. One of the most powerful spectroscopic techniques to determine the electron density is by the measurements of the Stark broadened line profile of an isolated atom or singly charged ion with reasonable accuracy [31, 32, 61, 62]. Stark broadening of special lines in plasmas results from collisions with charged species which Stark shifts the energy levels of atom/ion undergoing transition, resulting in both a broadening of the line width and a shift in the line centre. In dense plasmas with temperatures in the 1-10 eV range, the dominant line broadening mechanism is Stark broadening. Stark broadening is principally a density effect and does not depend sensitively on the temperature or on the electron velocity distribution. Stark broadening is thus essentially independent of the assumption of local thermodynamic equilibrium (LTE) [63, 64]. It follows that the Stark broadening measurements will give reliable electron densities even in cases where the existence of LTE is doubtful, whereas some other methods would then become invalid.

In Stark broadening method absolute photon intensities are not required, and merely the relative line shapes and widths are sufficient to calculate electron density. Here, full widths at half maximum (FWHM), ie., wavelength differences between two points where the intensity has fallen by a factor of 2 from that at the maximum, should be used. Their measurements require some relative intensity calibration or test of linearity of the detector, but background corrections are not essential, as long as intensity maximum of the line is well above this background. The FWHM of the lines $\Delta\lambda_{1/2}$ is related to the electron density by the expression [31]

$$\Delta\lambda_{1/2} = 2W\left[\frac{n_e}{10^{16}}\right] + 3.5A\left[\frac{n_e}{10^{16}}\right]^{1/4}\left[1 - \frac{3}{4}N_D^{-1/3}\right]W\left[\frac{n_e}{10^{16}}\right]\text{\AA} \qquad (1)$$

where W is the electron impact parameter which can be incorporated to different temperatures [31]; A, ion broadening parameter and N_D the number of particles in the Debye sphere defined as

$$N_D = 1.72 \times 10^{12}\frac{[T(eV)]^{3/2}}{[n_e(m^{-3})]^{1/2}} \qquad (2)$$

The first term in the right side of eqn. 1 represents the broadening due to electron contribution and the second term is the ion correction factor. For nonhydrogenic ions Stark broadening is predominantly by electron impact. Since the perturbations caused by ions is negligible compared to electrons, the ion correction factor can safely be neglected. Therefore eqn. 1 reduces to

$$\Delta\lambda_{1/2} = 2W\left[\frac{n_e}{10^{16}}\right]\text{\AA} \qquad (3)$$

In order to estimate the electron density, the Stark broadened profile of C II transition at 392 nm ($3p^2p^0 - 4s^2S$) is charted keeping the monochromator at its maximum resolution. Three broadening mechanisms are likely to contribute significantly to UV/visible line widths observed in plasmas produced during pulsed laser deposition, *viz.*, Doppler broadening, resonance pressure broadening and Stark broadening. For ablation in vacuum, where ablated species exhibit high expansion velocities, one of the dominant contributions to spectral line broadening is motional Doppler broadening (particularly at distance ≥ 5 mm from the target surface) which is due to different Doppler shifts (ie, $\Delta\lambda = \lambda v_z/c$) experienced by the species in different regions of the plume having different velocity components v_z in the observation direction. Typically for pulsed laser deposition plasmas the velocity component v_z is $\sim 10^6$ cm s^{-1} [9] which, for lines in the visible region, corresponds to Doppler line widths (FWHM) ~ 0.1 Å. The resonance pressure broadening interaction occurs on strong resonance lines of an atom/ion when excited atom/ion interacts with identical ground state atoms/ions. The resulting broadening is proportional to the ground state number density of the corresponding species and transition oscillator strength. Since the relatively small transition oscillator strength has been reported [65] for C II at 392 nm (0.134), the resonance broadening part can be safely neglected.

One of the simplest methods for the determination of electron temperature is by using relative intensities of lines of the same element and ionization state. This is one of the earliest method for the determination of temperatures in LTE plasmas, which is based on the fact that densities in various excited states are proportional to the product of statistical weights with the exponential of the negative ratios of excitation energy and the thermal energy kT. The temperature is accordingly inversely proportional to the logarithm of the ratio of the total intensities of lines arising from different upper levels, provided that none of these lines is affected by self absorption [27]. The electron temperature T_e can be determined using the eqn. [31]

$$kT = \frac{E' - E}{ln(I\lambda^3 g' f'/I'\lambda'^3 gf)} \tag{4}$$

where I, λ, g and f are the intensity, wavelength, statistical weight of the lower state and oscillator strength of the one line and E its excitation energy. Primed quantities represent corresponding constants for the second line.

Relative line intensities from the same element and ionization state usually do not provide accurate temperatures. The principal reason for this is the relatively small separation between the upper levels of two lines. This renders the line intensity ratio rather insensitive to temperature changes. Considerable improvement in sensitivity can be obtained by selecting lines from successive ionization states of the same element, because the effective energy difference is now enhanced by the ionization energy, which is larger than the thermal energy. In LTE, the ratio of such line intensities is given by [31],

$$\frac{I'}{I} = \left[\frac{f'g'\lambda^3}{fg\lambda'^3}\right]\left[4\pi^{3/2}a_0^3 n_e\right]^{-1}\left[\frac{kT_e}{E_H}\right]^{3/2} exp\left[-\frac{E' + E_\infty - E - \Delta E_\infty}{kT}\right] \tag{5}$$

where the primed symbols represent the line of the atom with higher ionization stage; a_0 is Bohr radius; E_H, ionization energy of the hydrogen atom and ΔE_∞ is the reduction to the ionization energy E_∞ of the lower ionization stage.

The correction factor in the ionization energy is given by

$$\Delta E_\infty = 3z\frac{e^2}{4\pi\epsilon_0}\left[\frac{4\pi n_e}{3}\right]^{1/3} \tag{6}$$

where z = 2 for the lower ionization state.

During the spectral studies of laser produced carbon plasma, we have observed ionized species up to C IV along with molecular species like C_2. We have not observed any line of considerable intensity corresponding to excited carbon atom (CI) in the wavelength region studied. Intensities of the successive ionized lines of carbon *viz*, C II, and C III are used for these calculations. For the evaluation of T_e we make use of line intensities corresponding to the lines at 464.7 nm and 569.6 nm of C III and 392 nm and 407.4 nm lines of C II. The spectroscopic constants of these spectral lines are given elsewhere [32]

3.1 Spatial dependence

The plasma represents a heated high pressure gas kept in a region of small dimensions which later on, is allowed sudden expansion into the surrounding vacuum. In a laser produced plasma the preferential vaporization of the evaporated material is always found to be in a direction perpendicular to the target surface, irrespective of the angle of incidence of laser beam. Investigation of optical emission in the neighbourhood of the target surface in the early stages of plasma evolution give direct information about the laser - target interaction as well as laser - plasma interaction [66]. Spectroscopic studies at comparatively larger distances from the target surface results in yielding information on the plasma species reactivity which is an important quantity needed to maintain quality of thin films prepared using the pulsed laser deposition techniques [67]. Line shape analyses were repeated at different distances from the target surface which provide a direct indication of space evolution of electron density giving an insight into the basic ionization processes taking place in the pulsed laser ablation. It is possible to distinguish between two phases in the time and space evolution of the laser induced vapour plasma. The first one is the ionizing phase characterized by laser heating of the target and the dense ionized vapour in the vicinity of the surface. The second one is the recombining phase of the expanding plasma [68, 69]. The first phase is not accessible for optical emission spectroscopic observation because of the reduced dimension of the dense ionizing vapour layer and its high optical opacity. The only indicator of the ionizing phase is the continuum emission which is very intense in the vicinity of the target surface.

The spatial dependence of electron temperature and density of the carbon plasma were carried out for distances up to 12 mm from the target surface and are given in figures 2 and 3. The spectra are charted at a pressure of 2×10^{-5} mbar and at an irradiance of 50 GW cm^{-2}. With increasing separation from the target surface, the electron temperature falls from 2.43 eV to 1.6 eV. The variation of electron temperature with distance (z) perpendicular to the target surface shows a $z^{-0.1}$ dependence. For these studies time integrated intensities were used and the value of T_e presented in different distances from the target should be regarded as indicative of the average conditions occurring in an Nd:YAG laser induced carbon plasma, rather than defining the conditions at a particular stage of its evolution.

The density of the plasma at a point z, at any time t can be expressed as [28]

$$n_e(z, t) = n_0(t) \frac{t}{\tau} \left(1 - \frac{z}{Z(t)} \right) \qquad (t < \tau) \qquad (7)$$

where n_0 is the density at the centre of the laser irradiated spot (z = 0) at time t; t/τ takes into account of the injection of particles into the plasma (τ, pulse width of the laser); the z coordinate is directed perpendicular to the target and Z(t) refer the spatial coordinate of the leading edge of the plasma. The above equation shows that the number of particles of the plasma increases linearly with time (for

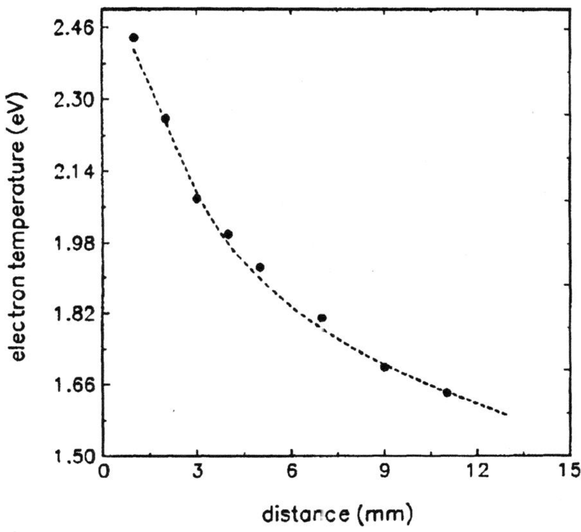

Figure 2: The variation of electron temperature as a function of distance from the target surface (laser irradiance used 50 GW cm^{-2} and pressure 2×10^{-5} mbar)

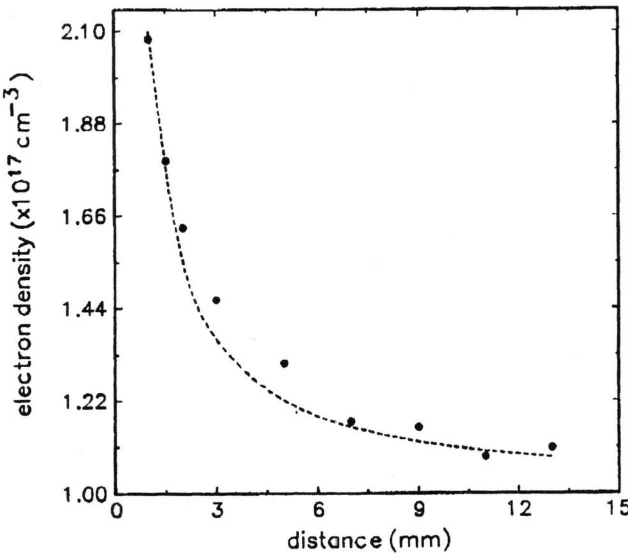

Figure 3: Electron density of the graphite plasma as a function of distance from the target surface (laser irradiance used 50 GW cm^{-2} and pressure 2×10^{-5} mbar)

$t < \tau$, isothermal expansion) during the incidence of the laser pulse due to constant evaporation from the target. In the adiabatic expansion there is no increase in the number of particles $(t > \tau)$, the density gradients can be expressed as [28]

$$n(z,t) = n_0(t)\left(1 - \frac{z}{Z(t)}\right) \qquad (t > \tau) \tag{8}$$

The spatial dependence of electron density of the carbon plasma shows the plasma expansion is characterised by a rapid decrease of density. With increasing separation from the target surface, the electron density decreases from $2.1 \times 10^{17} cm^{-3}$ at 1 mm to $1 \times 10^{17} cm^{-3}$ at 11 mm. The decrease of n_e as a function of distance is following approximately the $1/z$ law at short distances, indicating that the initial expansion of the electron gas is one dimensional, in good agreement with the predictions of the plume expansion model given by Singh and Narayan [29].

3.2   Time dependence

The temporal evolutions of electron temperature and electron density are of prime importance, since many kinetic reaction rates depend directly or indirectly on these parameters. The experimentally measured variations of temperature and density of the laser generated carbon plasma with time are given in figures 4 and 5 respectively. For these studies the boxcar gate width was set at 10 ns. Due to high density of particles in the initial stages of the plasma expansion, there are numerous collisions lead to free-free (bremsstrahlung) and free-bound (recombination) transitions and plasma behaves as a continuum of compressible fluid. At shorter times (< 100 ns), the line to continuum ratio is so small and the temperature measurement is very sensitive to errors in setting the true continuum level. This problem is particularly acute for times up to 50 ns. For time interval > 100 ns, the line to continuum ratios are within reasonable limit, interference with the continuum measurement is not severe and the vales of T_e shown in the figure 4 should be reliable.

Initially the plasma expands isothermally within the time of the duration of the laser pulse. After the termination of the laser pulse, the plasma expands adiabatically. During this expansion the thermal energy is converted into kinetic energy and the plasma cools down rapidly. An adiabatic expansion of the plasma occurs, when the temperature can be related to the dimensions of the plasma by the adiabatic thermodynamic relation

$$T[X(t)Y(t)Z(t)]^{\gamma-1} = constant \tag{9}$$

where γ is the ratio of specific heat capacities at constant pressure and volume, $X(t)$, $Y(t)$ and $Z(t)$ are the dimension of the expanding plasma in the three mutually orthogonal directions.

It is noted that within 300 ns after the laser pulse the temperature drops from 3.6 eV to 2.1 eV while the density falls from $3.5 \times 10^{17} cm^{-3}$ to $2 \times 10^{17} cm^{-3}$. In

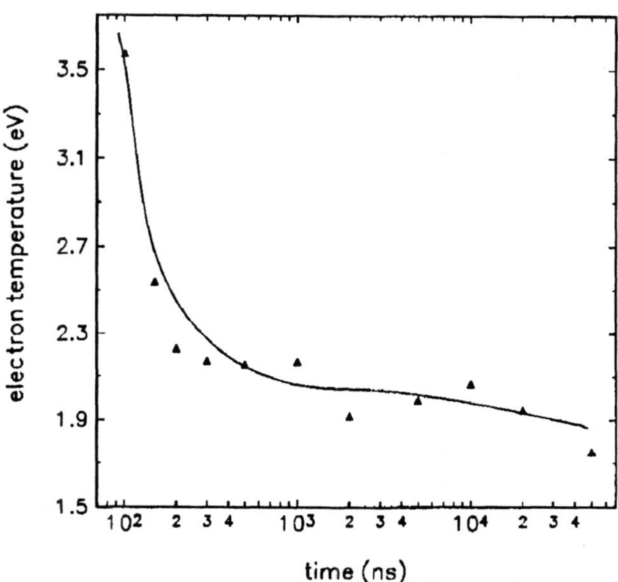

Figure 4: Electron temperature of the expanding plasma as a function of time (distance 3 mm; pressure 2×10^{-5} mbar, laser irradiance 50 GW cm^{-2})

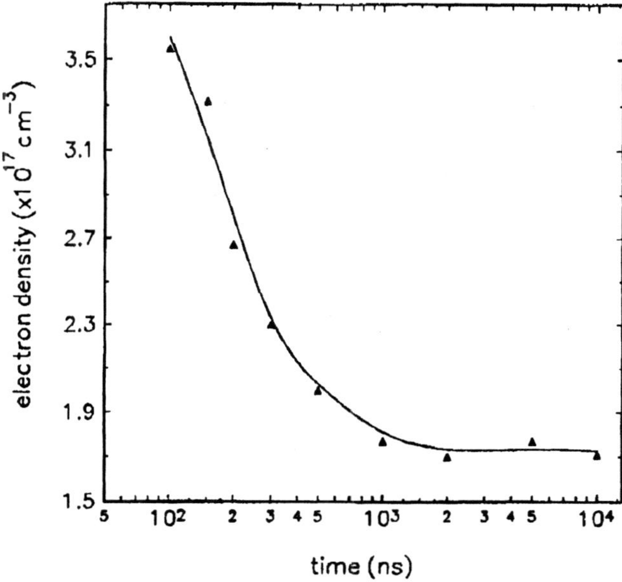

Figure 5: Electron density of the expanding plasma as a function of time (distance 3 mm; pressure 2×10^{-5} mbar, laser laser irradiance density 50 GW cm^{-2})

the early stage of plasma evolution the electron temperature is so high and it varies very rapidly. When the time is greater than 300 ns, the electron temperature of the plasma is reduced to about ~ 2 eV. But afterwards the temperature gets stabilized for a period up to ~ 2000 ns. As seen from the adiabatic equation of state, the rate of decrease of temperature strongly depends on the specific heat ratio γ. However, in actual practice the temperature decreases more slowly than predicted by the adiabatic equation due to preferential expansion of the plasma in one dimension during initial stages and due to recombination effects. But, 300 ns after the initiation of the plasma the temperature begins to decrease more slowly due to the energy released by the recombination which compensate the cooling due to expansion processes. The variation of electron temperature with time for 1.06 μm radiation show t^{-2} dependence which is in accordance with the theoretical adiabatic expansion model by Rumbsy and Paul [70].

The variation of n_e with time can be explained as follows. In the adiabatic regime, according to eqn. 8, the density decreases linearly with time. But in actual practice the density varies exponentially with time because of the high expansion velocities of the leading edge of the plasma which makes the plasma transparent the laser beam. The electron density has been found to decay with a dependence $n_e \propto t^{-2}$ rather than the reported theoretical $n_e \propto t^{-3}$ [70]. Such discrepancies from the t^{-3} has been noted by others also [70].

3.3 Dependence on Laser Irradiance

The nature and characteristics of the laser produced plasma strongly depend on the laser irradiance. For low intensities the vapour formed by the leading edge of the laser pulse never becomes sufficiently hot or dense to interact significantly with the laser beam. In this regime the main processes include heat conduction, melting and vaporization of the target. These thermal models are valid for irradiances up to the threshold of plasma formation. At high laser irradiance the vapour becomes significantly ionized absorbing part of the incident radiation and leading to vapour breakdown and plasma formation. As the plasma density and temperature increase, the plasma efficiently shields the target from the trailing part of the laser. Thus in this regime, the features of the laser ablation is strongly influenced by laser-plasma coupling and plasma kinetics.

Figures 6 and 7 give variation of electron temperature and density of the laser produced carbon plasma with respect to laser irradiance. As laser irradiance increases from 21 GW cm^{-2} to 64 GW cm^{-2}, the electron temperature increases from 1.29 eV to 2.15 eV, and saturates at higher irradiance levels while electron densities varies from $1.2 \times 10^{17} cm^{-3}$ to $1.6 \times 10^{17} cm^{-3}$ and then saturates. The saturation in T_e and n_e at higher irradiance conditions is expected to be due to plasma shielding, ie., absorption and/or reflection of the laser photons by the plasma itself [1]. The reflection of the incident laser beam depends on the plasma frequency ν_p, which should be lower than the laser frequency. For Nd:YAG laser with its fundamental

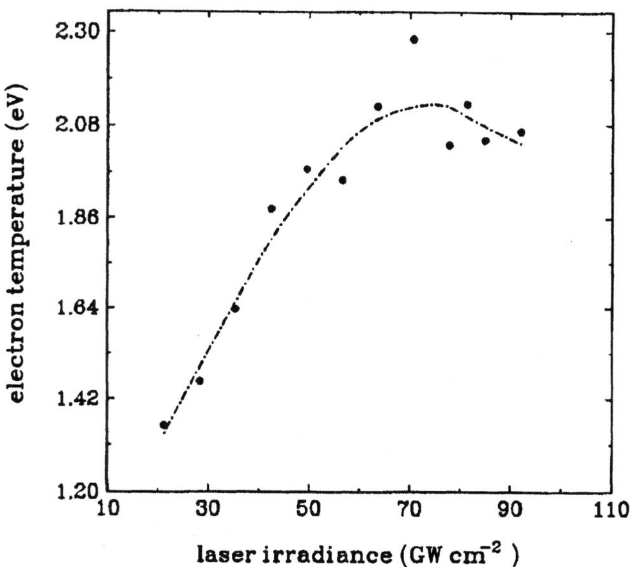

Figure 6: The variation of electron temperature with laser irradiance (distance 3 mm)

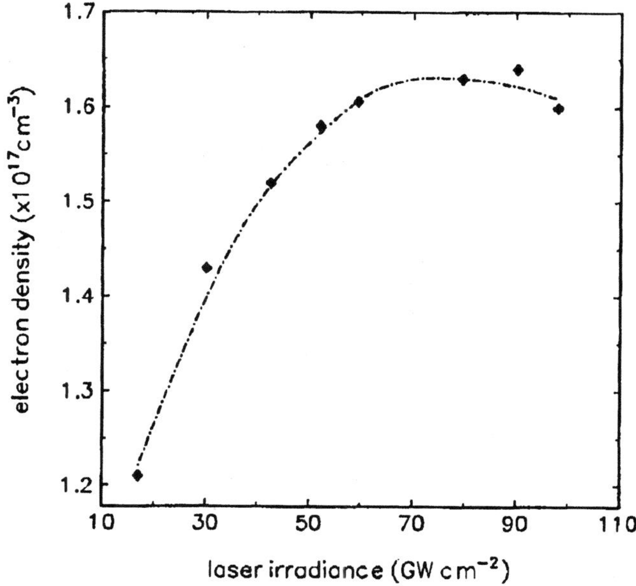

Figure 7: The variation of electron density with laser irradiance (distance 3 mm)

wavelength (1.06 μm) corresponds to a frequency $\nu_l = 2.828 \times 10^{14}$ Hz. The plasma frequency is given by $\nu_p = 8.9 \times 10^3 n_e^{0.5}$, where n_e is the electron density. Calculations show that, with $n_e \sim 10^{17} cm^{-3}$, $\nu_p = 6.5 \times 10^{12}$ Hz, which is much smaller than the laser frequency. So the energy losses due to reflection of Nd:YAG laser photons from the plasma can be assumed to be insignificant. An expression for the plasma absorption coefficient k_{pl} which depends on the plasma and laser light frequencies can be written as [71]

$$k_{pl} = \frac{2\nu_l}{c} Im \left[\left(1 - \frac{\nu_p^2}{\nu_l^2[1 + (i/\tau_{ei})]} \right)^{1/2} \right] \tag{10}$$

where τ_{ei} is the electron-ion collision time. The above equation is derived considering electromagnetic wave propagation in plasma, and is valid when the quasistationary approximation is satisfied, ie, when the amplitude of the electric field varies slowly with time. Since, ν_l, in our case, is much larger than ν_p, this approximation is valid. From this expression it follows that the absorption coefficient for laser radiation is approximately proportional to the square of the electron density, the square of the light wavelength and to $T_e^{-3/2}$. Essentially the same result is obtained by treating absorption as due to inverse bremsstrahlung [72]. This expression also reveals that the plasma absorption coefficient has a pronounced maximum when the laser frequency is in the vicinity of the plasma frequency.

When the vapour initially becomes slightly ionized, it absorbs part of the incident laser radiation. This energy is converted into internal energy of the plasma. With increase in plasma temperature, the degree of ionization of the vapour is also increased, thereby further enhancing absorption of the incident laser radiation. As the plasma density and temperature rise, the vapour phase progressively behaves like an optically thick medium. The plasma effectively shields the target surface from the trailing part of the laser pulse. The energy in the plasma is then transferred away by thermal radiation or converted into hydrodynamic motion. Due to lateral expansion of the plasma, the energy radiated from the plasma to the target is often distributed over an area significantly larger than the laser beam spot.

A radiating gas may be composed of molecules, atoms, ions and electrons. These particles have various energy levels associated with them. In accordance with the general scheme of allowed energy states of an atomic system, the electronic transitions accompanying absorption and emission of light are subdivided into three types: free-free, bound-free and bound-bound transitions. From an energy point of view on continuous spectra, bound-free and free-free transitions are of primary interest. The reason is that the radiant energy losses by distinct spectral lines usually represent small part of the continuous spectrum [72]. In this study it is assumed that the evaporated particles are either in atomic form or ionized states. The two dominant mechanisms responsible for plasma absorption at these laser irradiance levels used in our work are inverse bremsstrahlung and photoionization. Inverse bremsstrahlung absorption α_{ib} via free electrons is approximated by [72]

$$\alpha_{ib}(cm^{-1}) = 1.37 \times 10^{-35}\lambda^3 n_e^2 T_e^{1/2} \tag{11}$$

where λ is the wavelength of the laser photons in μm. It is noted that the absorption due to inverse bremsstrahlung is negligibly small at low irradiance levels and increases with increasing laser irradiance [7].

Absorption via photoionization can be estimated with Kramer's formula and absorption coefficient [72, 73]

$$\alpha_{pi} = \sum_n 7.9 \times 10^{18}\left[\frac{E_n}{h\nu_l}\right]^3\left[\frac{I}{E_n}\right]^{1/2} n_n \tag{12}$$

where E_n and n_n are the ionization energy and number density of the excited state n; h, Plank's constant; ν_l, laser frequency; I, ionization potential of the ground state atom. The absorption coefficient of photo ionization is obtained by summing up all the excited states whose ionization energy are smaller than the laser photon energy. Since the excitation potentials of atomic transitions of carbon atoms are much greater than the photon energy of the pump used (1.17 eV), direct photoionization by the absorption of a laser photon is ruled out. The only possibility for this type of ionization to happen is by the simultaneous absorption of number of photons.

The ionization rate in the case of photoionization is given by [22]

$$W_n = \omega_0 n_e^{3/2}\left[\frac{\xi_{os}}{I_i}\right]^p \tag{13}$$

where $p = I_i/\hbar\omega_0$ is the number of photons absorbed, I_i is the ionization energy and ξ_{os} is the electron oscillation energy which is given by

$$\xi_{os} = 0.093\lambda^2 I(eV) \tag{14}$$

where I_i is the power density used for the ablation (W cm^{-2}). Since the ionization rate depends on number of absorbed quanta p and the laser irradiance through ξ_{os}, the slope of the $\log - \log$ plot between laser irradiance and emission intensity will be a direct measure of number of photons involved in this process. We have seen that the intensity of the line grows as I^p where p = 10 for C II which is in good agreement with the photon energy used in the present experiment and the first ionization potential of carbon atoms (11.3 eV). But the value of p = 9 is observed for C III, implies that along with photoionization, other processes like impact ionization and direct generation of the species at higher energy states also take place [9]. At higher irradiance levels the exponent is found close to unity and obviously the multiphoton ionization is not a dominant process in this regime. Besides photoionization, other mechanisms especially impact ionization, thermal ionization etc. may also affect the absorption coefficient of the plasma.

The saturation in T_e and n_e at higher irradiances cannot be explained by considering only the prominent absorption mechanisms via, inverse bremsstrahlung and photoionization. Such temperature behaviour can be explained by assuming the formation of a self-regulating regime at higher irradiances. During isothermal expansion a dynamical equilibrium exists between the plasma absorption coefficient and rapid transfer of thermal energy into kinetic energy, which controls the isothermal temperature of the plasma. At high irradiance levels, when an appreciable amount of energy is absorbed by the plasma, a self-regulating regime may form near the target surface. If the absorption of the laser photons by the plasma becomes higher due to high plasma density, the evaporation of the species from the target becomes less, which in turn decreases density of the charged species. This consequently increases the absorption of the laser photons by the target, which in turn increases the temperature of the plasma. On the other hand, when the absorption of the laser energy is less, the process is reversed, with similar results. It has been theoretically proved that the density, temperature and dimensions of the plume adjust in such a manner that the plasma absorbs same amount of laser radiation to maintain a self-regulating regime [74]. This assumption was found to be valid in laser generated plasma, where thermalization time is significantly less than the plasma expansion time, resulting in an establishment of uniform temperature in the plasma. The thermalization time τ_{ei} of energy exchange between electron and ions/neutrals during collision can be estimated from the relation [75]

$$\tau_{ei} = \frac{252 M T_e^{3/2}}{\ln(\Lambda) n_e} \tag{15}$$

where

$$\Lambda = \frac{3(kT)^{3/2}}{4(\pi n_e)^{1/2} e^3} \tag{16}$$

where $\ln \Lambda$ stands for the Coulomb logarithm, which involves dynamical information about ion-electron collisions, M is atomic weight. With $n_e = 10^{17}$ cm^{-3}, $T_e \approx T$ (vapour temperature) = 22000 K, the relaxation time is $\sim$ 3 fs, which is much smaller than the expansion time or pulse width of the laser beam which is of the order of few of nanoseconds.

3.4 Pressure dependence

The study of plasma formed during the interaction between a laser beam and solid target in an ambient gas is of great interest [3, 76, 77, 78, 79, 80]. Experimental observation have shown that significant differences between plume expansion in vacuum and in the presence of a background gas. Recently a theoretical model has been made for the expansion of laser generated plasma into a background gas based on a combination of multiple-scattering and hydrodynamic approaches [16, 81]. Laser induced fluorescence studies [82, 83] have indicated that the ambient atmosphere acted not only as a buffer to increase the residence time of the ablated atoms, but also helped to atomize ablated droplets. Some reports have also discussed the effect of ambient atmosphere on the improvement of the analytical performance of atomic emission spectrometry using laser induced plasmas [84, 85]. Iida [86] investigated the atmospheric effects on the laser vaporization process. Time-resolved spectroscopy and direct measurement of vaporized sample mass revealed that the atmosphere present during the vaporization phase was very important for both the effectiveness of the vaporization and the subsequent plasma emission measurement. Ownes and Majidi [87] studied the effects of increased pressure of buffer gas on the laser-induced plasma emission. The three different laser intensity thresholds for plasma initiation in the presence of an ambient gas can be distinguished as the surface vaporization threshold, the target vapour plasma threshold and the ambient gas plasma threshold. These threshold values for vaporization, vapour plasma and gas breakdown strongly depend on experimental parameters like target material, wavelength of the laser radiation, nature of the ambient gas etc.

Compared to the expansion into vacuum, the interaction of the plume with an ambient gas is a far more complicated gas dynamic process which involves deceleration, attenuation, thermalization of the ablated species, and the formation of the shock waves. Due to the complexity of the ablation dynamics, an appropriate theoretical description of plume expansion for a wide range of ablation conditions is lacking. In low gas pressure, the plume propagation could be described by Monte Carlo Simulation [78, 88]. In moderate or high pressures, which is typical for thin film preparation, a blast wave model is found to describe accurately the plume propagation distance during the early expansion stages, whereas a shock layer model and an empirical drag model predict the maximum plume length with considerable accuracy [89, 90].

Studies were made by keeping the graphite target in three different atmospheres *viz.* air, helium and argon from 1 mbar to a pressure reduced to 10^{-5} mbar. The emission spectra, electron temperature and density are found to be significantly influenced by the ambient atmosphere. The addition of an ambient gas enhances the emission from all species. The relative enhancement depends on the gas pressure, the nature of the gas, and also the excitation energy of the electronic transition responsible of the line. The lifetime of all the transitions studied is of the order of few nanoseconds, while the observed increase of the emission occurs at a time in the order of few tens of nanoseconds [91], the increase in intensity should be due to species that have been excited during the plasma expansion. The two main processes invoked in the excitation are the particle collision excitation and electron impact excitation. Figures 8 and 9 give the variation of time integrated electron temperature and electron density of the carbon plasma with pressure at different ambient atmosphere (Air, He and Ar). It is noted that the electron density shows a decreasing trend with an increasing pressure irrespective of the nature of the ambient gas used while the electron temperature shows a somewhat different behaviour with respect to nature and pressure of the ambient gas.

The shielding effect, ie. the absorption of the laser energy by the plasma strongly depends on nature of the background gas used. Ambient gas breakdown will profoundly influence the laser energy coupling to the target surface. If gas breakdown occurs before laser light reaches the sample surface, major part of the energy will be absorbed by the resulting plasma formed from the gas. The laser irradiance used in these studies is 50 GW cm^{-2}. The breakdown thresholds of helium and argon reported using Q-switched ruby laser are 3×10^{11} and $1 \times 10^{11} W cm^{-2}$ at atmospheric pressure and increase gradually with decrease in gas pressure [92]. In the present experimental work, the pressure range used for Ar and He are well below atmosphere pressure and one may safely predict that gas breakdown will not occur.

A cascade growth of the electron number density and absorption coefficient of the plasma will be greatly influenced by the nature of the background gas. A cascade like growth of electron density is an important process because of the absorption of the laser radiation by the plasma with increase in ambient gas pressure. The condition necessary for the development of cascade like growth is given by [93]

$$\frac{d\epsilon}{dt} = \frac{4\pi^2 e^2 I \nu_{eff}}{m_e c \omega^2} - \frac{2 m_e \nu_{eff} E}{M} > 0 \qquad (17)$$

where ϵ is the energy of the free electrons; e and m are the charge and mass of the electron; M, mass of the background gas neutral particle; E, the energy of the first ionization stage of the gas; ν_{eff}, the effective frequency of $\bar{e}$ - neutral collision; I, radiation intensity and ω, the cyclic frequency of radiation. The first term on the right hand side of the eqn. 17 expresses the rate of growth of energy by the absorption of laser photons, and the second term gives the maximum rate of energy loss due to elastic and inelastic collisions with neutral gas particles.

According to the classical electrodynamics, the spectral absorption coefficient (k_{ib}) due to inverse bremsstrahlung is given by [94]

$$k_{ib} = [1 - exp(-\hbar\omega/kT)]\left[\frac{16e^6\pi^2}{3\hbar c m_e \omega^3}\right]\left[\frac{2\pi}{3m_e kT}\right]^{1/2} n_e \sum_i Z_i^2 n_i g_i \qquad (18)$$

where k is the Boltzmann constant, h, Planck's constant, T temperature, n_i and n_e represent the number of ions and electrons per unit volume, Z_i stands for the charge of ions and g_i, the Gaunt factor which corrects the semiclassical expression for quantum effects. In the above expression, the inverse bremsstrahlung contribution stemming from electron-neutral interaction has been neglected. This equation indicates that higher the electron density and degree of ionization, the more effectively the laser radiation is absorbed by the plasma.

When the plasma medium absorbs a significant fraction of laser energy, a laser supported detonation wave will be formed. Under this condition the length of the

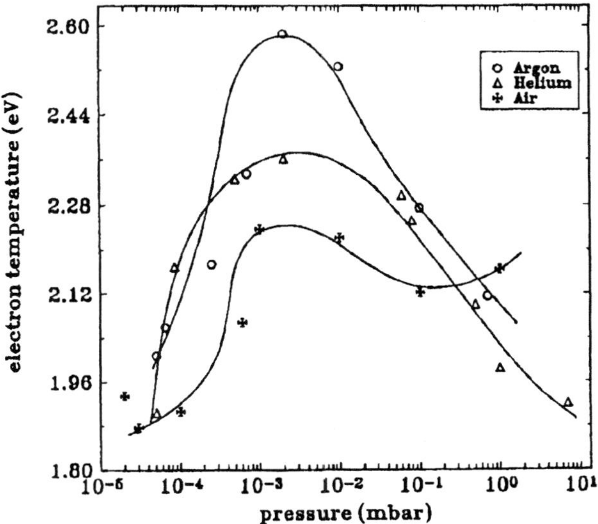

Figure 8: Variation of electron temperature as a function of pressure of the different ambient gas used (distance 3mm)

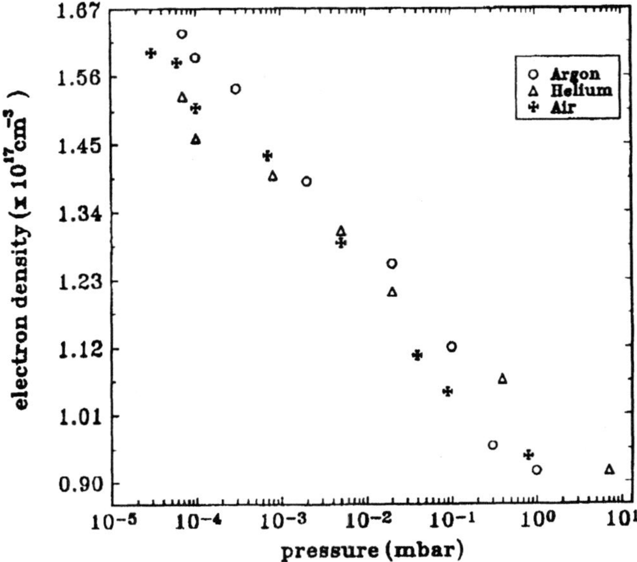

Figure 9: Variation of electron density as a function of pressure of the different ambient gas used (distance 3mm)

plasma can be written as [95]

$$z = const\left(\frac{E_F^2\sigma(\gamma^2 - 1)}{\rho\beta}\right)^{1/5} t^{3/5} \tag{19}$$

where E_F is the electric field strength; γ, specific heat ratio; ρ, density of the ambient gas; β an absorption factor expressed partly by eqn. 18; and σ, the high frequency conductivity which is given by

$$\sigma = \frac{e^2 n_e \nu}{m_e(\nu^2 + \omega^2)} \tag{20}$$

where ν is the electron collision frequency.

At moderately high pressures, the plasma expansion can be modelled as a blast wave, which is represented by

$$z = \left(\frac{W}{\rho}\right)^{1/5} t^{2/5} \tag{21}$$

where W is the total energy absorbed. The dimension of the plasma, which is closely related to the density, temperature and coupling of laser radiation is influenced by the surrounding gas through eqns. 19 and 21. In our present work, the axial spread length of the plasma is considerably larger in helium atmosphere in comparison of argon and air, since density of helium is less than that of air and argon. To get a clear idea about the background gas effect on the nature of the plasma, the physical processes of the ambient gases should be studied in detail. Some physical properties of the gases, which are related to plasma generation processes is given in the table 1.

Table 1: Physical properties of different ambient gases used [96]

	He	Ar	Air(N_2)
Atomic/molecular weight	4	40	28
Ionization energy (eV)	24.6	15.8	15.6
γ	1.67	1.67	1.40
Density (at 760 torr)(g/litre)	0.1785	1.784	1.251
Thermal conductivity ($10^{-6}cals^{-1}deg^{-1}cm^{-1}$)	360.36	42.57	62.4

In comparing argon and helium atmospheres the cascade condition (eqn. 17) is more favoured for argon (M = 40, E = 14.5 eV) than He (M = 4, E = 23.4 eV). The plasma formed in Ar atmosphere are more absorptive than He atmosphere which directly influence the value of T_e and n_e. We have observed that the temperature of carbon plasma is higher in argon compared to the presence of helium or air. It can be seen that the electron temperature attains a maximum value at $\sim 10^{-2}$ mbar for argon and He ambient atmospheres, then decreases with change of pressure. But density is found to decrease with increasing pressure of the ambient gas irrespective of the nature of the buffer. As the pressure increases the confinement of the plasma nearer to the target surface takes place, which in turn increases the effective frequency of electron collisions with background gas atoms. The cascade condition is not favoured at high pressures because of the energy loss due to elastic collision of the electrons with the neutral particles of the gas which supersedes the rate of growth of energy of free electrons via inverse bremsstrahlung. The decrease in electron temperature at pressures $< 10^{-2}$ mbar is presumably due to less efficient confinement of the plasma, so that plasma energy is distributed over a large volume. As the pressure of the ambient gas increases, confinement of the plasma takes place, which improves the elastic and inelastic collisions and thereby recombination processes. This is supported by the fact that the increase in pressure decreases the electron density. The increase in temperature with pressure is therefore due to the energy gained by the recombination superseding the cooling due to increased intraplume collisions.

The temperature profile in air is different from the profiles in helium and argon particularly at higher pressures ($\geq 10^{-2}$ mbar). This means that T_e can be affected by chemical reactions in the plasma as well as by simple hydrodynamic expansion of the plasma. Higher temperature was obtained in 1 mbar air than was obtained with a similar argon and helium atmospheres. The nitrogen and oxygen present in air atmosphere influences the temperature through exothermic reactions[11]. The rate of change of electron temperature in the plasma is the sum of three terms *viz.*, elastic collision, electron heating due to collisional deexcitation of metastable ions and recombination of ions. The rate of loss of electron energy at short times is mainly dominated by elastic collision term $Q_{\Delta t}$, given by [70]

$$Q_{\Delta t} = \frac{2m_e}{M_B}\sigma_{ea}n_B\left[\frac{8kT_e}{\pi m_e}\right]^{1/2} \tag{22}$$

where σ_{ea} represents elastic scattering cross-section between electrons and atoms; n_B and M_B are the density and mass of background gas atom. It is clear from the eqn. 22 that, the cooling is inversely proportional to M_B and hence lighter gases are efficient for rapid cooling. Helium being the lighter gas compared to argon or N_2 (air) gives rise to rapid cooling as is observed in the present experiment.

4   TOF studies in carbon plasma

TOF studies of the plasma give vital information regarding the time taken by a particular state of the constituent to evolve after the plasma is formed. This technique gives details on the velocity of the emitted particles, parameters which are of fundamental importance in establishing the mechanisms responsible for the particle emission. Thus this measurement is important for obtaining knowledge of complicated ablation and transport processes occurring in the carbon plasma produced by the laser ablation technique. The time resolved studies of emission lines from various species originating from carbon plasma are carried out from the oscilloscope traces which show definite time delays for emission with respect to the incidence of the laser pulse. Our observations show that the emission lines from different emitting species within the plasma possess distinctly different temporal profiles. Each temporal profile represents a complex convolution of different factors that govern the temporal history of the emitting species *viz.* its production mechanism and rate, its flight past the viewing region and its radiative and collisional decay rates. In this section, a comprehensive study of spatial characteristic emission from C_2 using time resolved spectroscopy is given. Such temporally and spatially resolved high resolution spectroscopic studies are helpful to optimize parameters of DLC film deposition and to correlate the carbon clusters with plasma dynamics.

The time evolution of the spectral emission obtained in the present work clearly reveals that the C_2 species ejected from carbon target has a twin peak distribution up to a certain distance from the target (16 mm) and at farther distances the TOF pattern shows a triple peak structure. This interesting feature in TOF pattern of C_2 species becomes prominent only beyond a threshold laser irradiance. Below this threshold laser irradiance only single peak distribution is observed. Typical TOF distributions of C_2 species obtained by monitoring the spectral emission from C_2 in electronic excited state (at $\lambda = 516.5$ nm corresponding to the (0, 0) transition $d^3\Pi_g \rightarrow a^3\Pi_u$ of C_2 Swan system) at a laser irradiance of 81 GW cm^{-2} for different axial distances from the target are shown in the figures 10(a-f). The initial spike in the figure is due to scattering and/or prompt emission and can be used as a time marker.

The time resolved observation presented here characterizes the axial expansion of the plasma i.e, strictly along a direction perpendicular to the target surface. Figures 10 (a) and (b) represent the oscilloscope traces of TOF distribution of C_2 molecules at distances 5 mm and 10 mm from the target. At these distances, there exists a double peak structure. The emergence of the new peak occurs only at distances greater than 16 mm from the target. Fig. 10 (c) shows the formation of the new peak in between the aforesaid twin peaks. Figs. 10 (d-f) show the triple fold TOF distribution of the C_2 species at distances 18 mm, 20 mm and 22 mm away from the target surface. In what follows, for simplicity, the first or faster peak is designated as Pk1, delayed or slower peak as Pk2 and newly generated peak after spatial distance of 16 mm as Pk3.

Variation of time delay with distance for Pk1 and Pk2 are given in figures 11 and 12. It is seen from fig. 11 that the time delay for the faster peak is constant up to 10 mm from the target and then increases. It is also noted that after 17 mm the delays for Pk1 decreases sharply. TOF distributions give the time of arrival at a certain point in space with a known flight length and these can easily be transformed into velocity distribution. From the plots of delay time vs distance, one can obtain instantaneous velocities for C_2 molecules and these are given in figs. 13 and 14

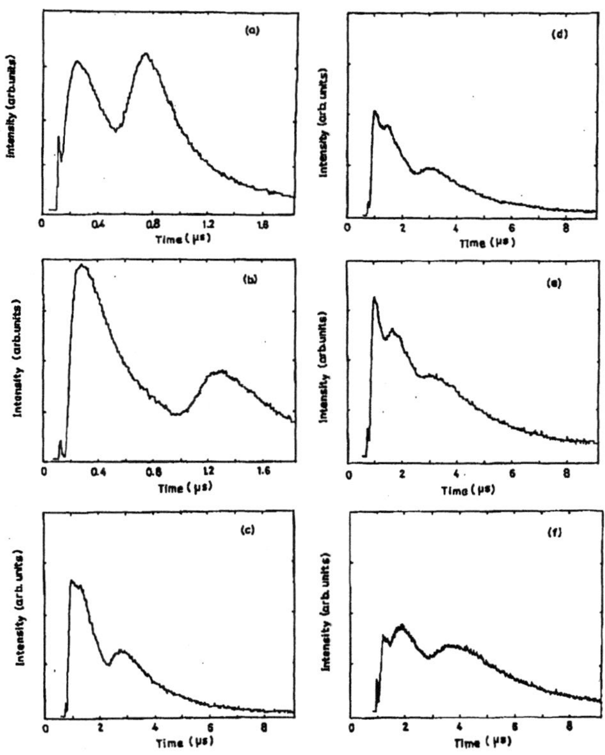

Figure 10: Intensity variation of spectral emission with time for C_2 species (516.5 nm) at different distances from the target. Distances are (a) 5 mm, (b) 10 mm, (c) 17 mm, (d) 18 mm, (e) 20 mm and (f) 22 mm. These TOF spectra are recorded at a laser irradiance 81.34 GW cm^{-2}

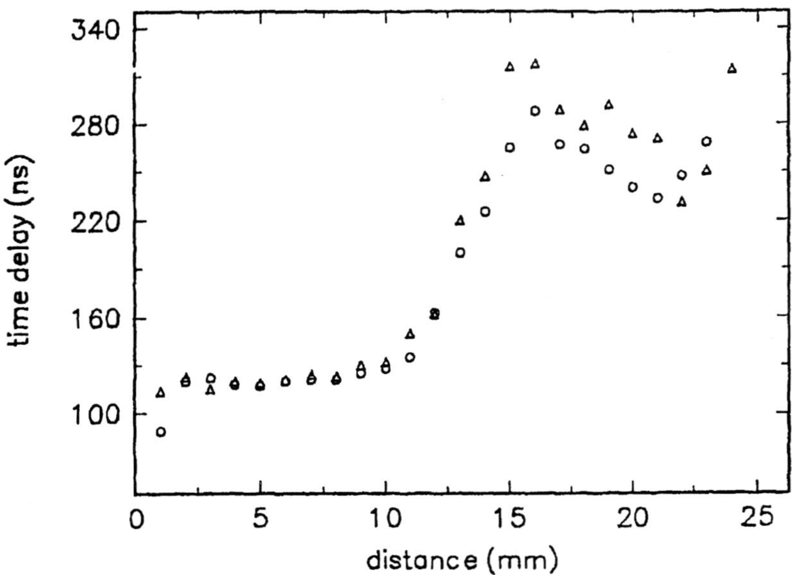

Figure 11: Variation of time delay in the peak intensities with distance for the Pk1 of C_2 at different laser irradiances. (o) 81.34 GW cm^{-2} and (Δ) 88.4 GW cm^{-2}

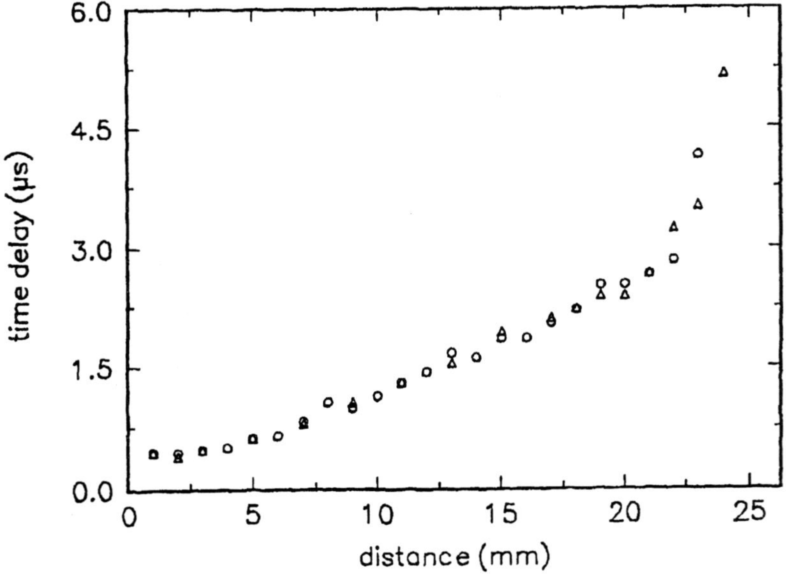

Figure 12: Variation of time delay in the peak intensities with distance for Pk2 of C_2 at different laser irradiances (o) 81.34 GW cm^{-2} and (Δ) 88.4 GW cm^{-2}

respectively. It may be noted that the velocities are not constant and they vary with distance from the target. From the mean velocity distribution of these species it is clear that the velocity of Pk1 increases with spatial separation from the target up to z = 10 mm. The sudden decrease in the velocity of these species after 10 mm shows the deceleration of the C_2 species. However at distances greater than z = 15 mm, the velocity of the particles again gets enhanced. In the case of Pk2 (figure 14) the velocity increases with spatial separation from the target until it reaches 6 mm and then the expansion velocity is found to be somewhat constant (8 Km s^{-1}) up to z = 20 mm and then decreases. The intensity variation of these peaks with spatial separation shows different spatial maxima for Pk1 (z = 12 mm) and Pk2 (5 mm). The variation of time delay for newly generated peak (Pk3) beyond z = 17 mm showed an increasing tendency with increasing distance.

The above observations indicate that the plasma apparently develops fast and slow components above a threshold laser irradiance. There are only a few reports available in the literature which describe multiple peak TOF distribution in laser generated plasma from graphite target. Abhilasha et. al [97] reported a peculiar double peak structure in the temporal profile of C II species in a laser produced carbon plasma at various air pressures and they proposed that, this may be due to stratification of the plasma into fast and slow ion components at the interface where the occurrence of Rayleigh - Taylor instability causes deceleration of the plasma front by ambient gas. Dixon and Seely [98, 99] also observed a double peak structure in C II species and they explain it as due to collisional interaction like resonance charge transfer which has been shown to be a velocity dependent one. In fact the existence of twin peak structure for the C_2 species from the laser induced plasma observed previously has been explained as due to emission from the 'shell' (fast) and the 'core' (slow) components [100]. Attributing peaks Pk1 and Pk2 solely to fast and slow components will not explain the unusual spatial dependence of time delays as depicted in figures 11 and 12 respectively. We have also observed a twin peak structure for Y I and YO emission in the laser produced plasma from $YBa_2Cu_3O_7$ and the twin peaks were assigned as due to species corresponding to those generated directly/in the vicinity of the target surface and to those generated from collisional and recombination processes [13]. Lowndes et al [35] recently observed three 'modes' of incident species in the TOF profile using ion probe method and they attributed it to scattered ions, ions that are slowed by gas phase collisions and slow moving clusters formed through collisions respectively. Tasaka et al [101, 102] observed triple fold plume structure during Nd:YAG laser ablation of graphite in helium ambient atmosphere. During optical emission studies using streak camera they found that, the fastest component is composed of carbon ions, second fastest component is due to compressed neutral molecules and the slowest component is the radial vapour from the graphite target. Bulgakov and Bulgakova [103] made a theoretical model for the plume expansion into ambient gas and have shown that the pulsating character in the velocity distribution can be explained using back and forth shock wave

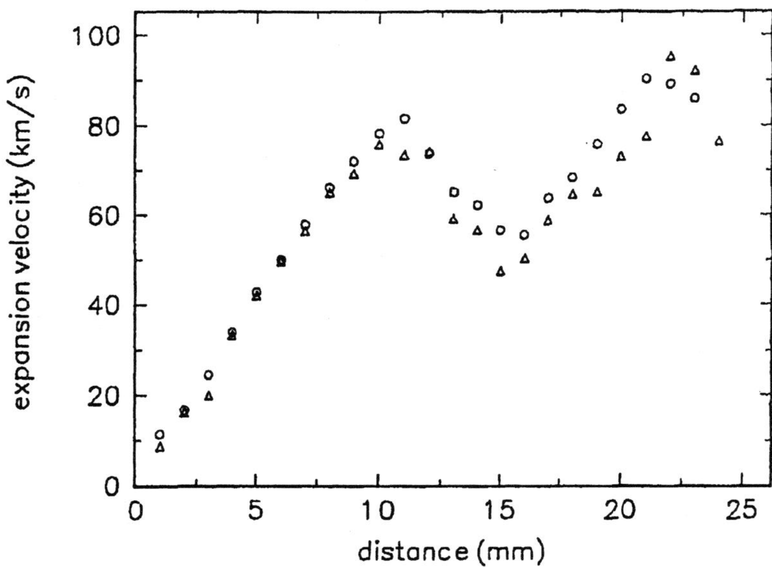

Figure 13: Expansion velocities as a function of distance for Pk1 of C_2 for laser irradiance (o) - 81.34 GW cm^{-2} and (Δ) - 88.4 GW cm^{-2}

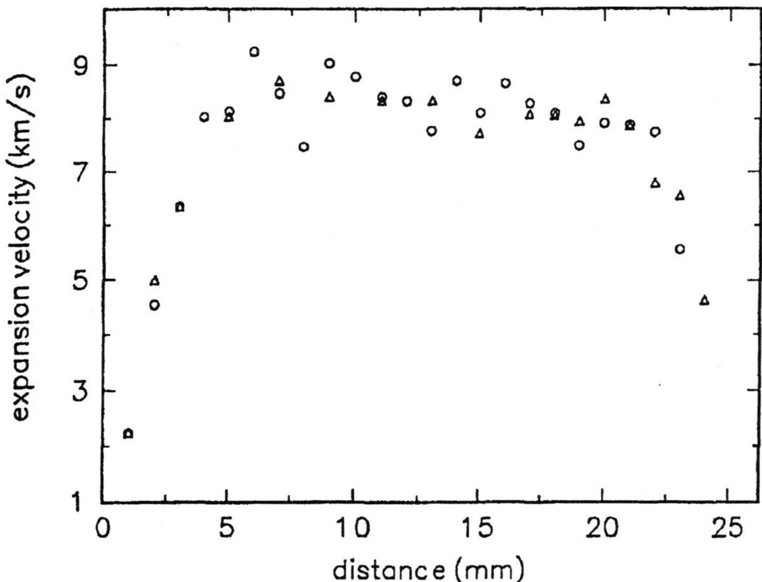

Figure 14: Expansion velocities as a function of distance for Pk2 of C_2 for laser irradiance (o) - 81.34 GW cm^{-2} and (Δ) 88.4 GW cm^{-2}

propagation and they also indicated that ionization and recombination processes have no significant effect on these pulsations. But they could not succeed in explaining the triple structure for BaO molecule during mass spectroscopic studies [104] using cloud ionization model.

The variation of time delay with laser irradiance for Pk1 and Pk2 are given in figs. 15 and 16 respectively. From these figures one can note that the delay of Pk1 increases with laser irradiance while that of Pk2 decreases sharply after being constant up to a certain value of irradiances. We also observed that the intensity of Pk1 increases from zero and saturates at higher irradiances while that of Pk2 increases up to an optimum laser irradiance and begins to decrease thereafter. This behaviour is predominant at points having smaller separation from the target.

There are two mechanisms that could exist for the particle formation during the laser ablation of carbon, *viz.*, dissociative and recombinational. Harano et al [105] have suggested the strip-off mechanism for the formation of carbon clusters by considering the balance between the total mass of the carbon particles and the laser irradiance used for the ablation. Most of the models of fullerene formation are based on the recombination mechanism, i.e, the formation by nucleation from carbon vapour consisting of carbon atoms and very small carbon molecules [106]. It is well known that graphite exhibits a large difference between the inter-layer and the intra-layer bond strengths. It is expected that at low laser irradiance levels, graphite will be ablated layer by layer producing large particles which in turn get dissociated to form C_2 species [60]. The dissociative mechanism can further be supported by the observation of long duration of Swan band emission at low irradiances. At low laser irradiance levels only a slowly propagating component with low kinetic energy is observed. The larger mass of C_n will result in longer delays which are observed in the C_2 emission (Pk2) occurring at the lower laser irradiances. Above a distance $\simeq$ 6 mm from the target, the plasma is colder compared with the same in core region and collisional effects become insufficient to cause dissociation of C_n. Therefore the dominant mechanism for the production of C_2 Swan band emission at low irradiances is the electron collision with C_n cations and neutrals (n >2) followed by dissociation where one of the fragments is an ejected C_2 molecule. As the laser irradiance is increased, clusters with lower values of n will be ejected directly from the target. Above a threshold laser irradiance, temperature of the plasma becomes so large so as to dissociate C_n to neutral and ionized carbon atoms just outside the target. So at higher irradiance levels Swan band formation is mainly due to electron-ion and ion-ion recombination. It is observed here that at high laser irradiances, after a threshold, an emission peak showing a faster component with higher kinetic energy for C_2 molecules begins to appear.

Spectral analyses show that at low irradiance levels, the emission spectrum is dominated by C_2 molecules whereas at higher irradiance levels the plasma emission is mainly due to ionic species of carbon up to C IV along with C_2 species. Emission originating from ionic species appears when the laser irradiance is sufficient to create

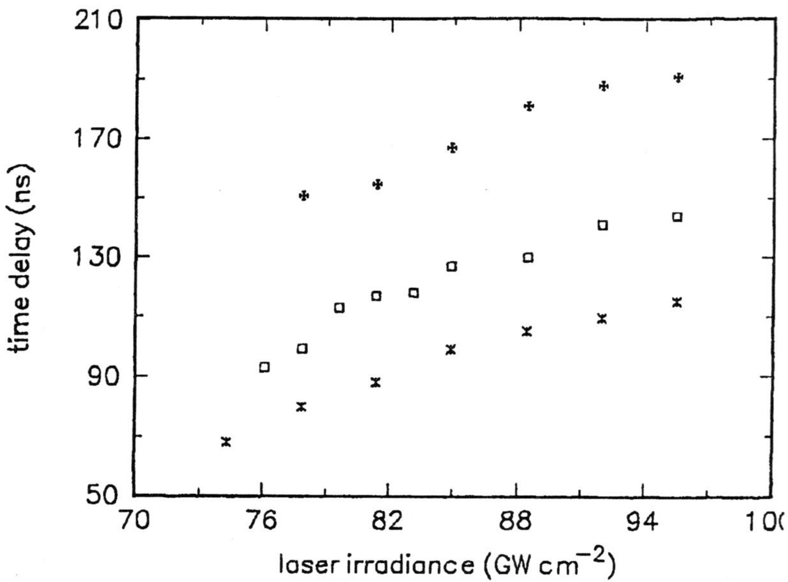

Figure 15: Plot of time delays on laser irradiance of Pk1 for different distances ($\star$) 5 mm, (box) 10 mm and (+) 15 mm

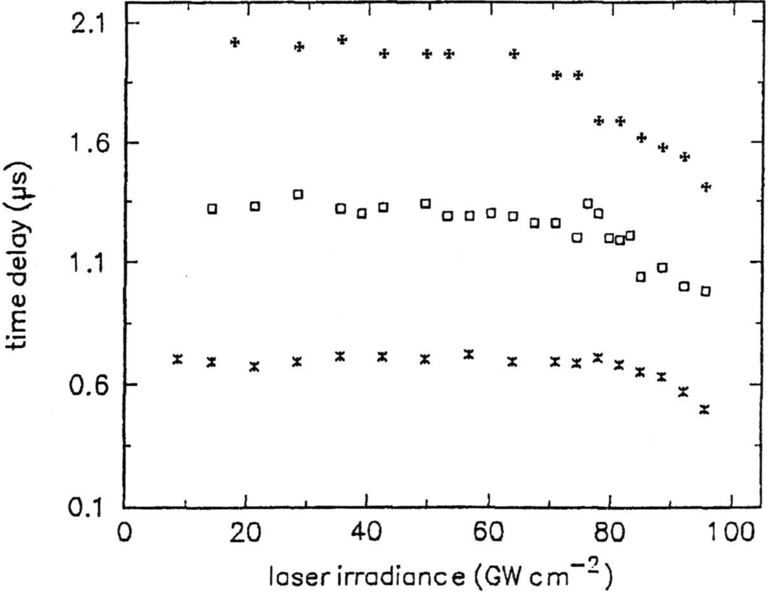

Figure 16: Plot of time delays on laser irradiance of Pk2 for different distances ($\star$) 5 mm, (box) 10 mm and (+) 15 mm

a predominantly ionized plasma medium. Temporal profiles are recorded for the ionized carbon species for different distances from the target. The ionic species are characterized by faster and narrower TOF distributions in comparison with atomic or molecular species. The expansion velocities of the ionized species are found to be increasing with degree of ionization. It is noted that the maximum expansion velocities of C II, C III and C IV are found to be at 40 Km/s, 58 Km/s and 80 Km/s respectively. The maximum spatial range for the ions were limited by the exponential drop in recorded intensity with distance. There is a laser intensity threshold for the appearance of multiply charged ions and this threshold increases with degree of ionization. It is also noted that the expansion velocities of these ionic species increase with increasing laser irradiance. The exponential increase in the intensity of emission of positive ions with irradiance is in accordance with the Richardson-Saha laws [107].

The observation of higher velocities for ions as compared to neutrals and molecules is due to Coulomb fields which are generated by negatively charged electrons escaping from the plume. Once ions and electrons are produced, one can have neutral carbon atoms by three body collision processes like

$$C^+ + e + e \rightarrow C + e \tag{23}$$

Collisional ionization is also possible through processes like

$$C + e \rightarrow C^+ + e + e \tag{24}$$

In the vicinity of the target, (24) may be predominant over (25) so that we get excited state C_2 formation slightly away from the target, giving emission peak Pk1. The threshold like phenomenon in the case of Pk1 also shows the initiation of production of ionized species so as to open the channels (24) and (25) at higher laser irradiances. As the laser intensity is increased beyond a threshold, the probability of cluster formation of the type C_n ($n > 2$) in the plasma diminishes, thereby causing a drastic decrease in the intensity of Pk2. Moreover, as the laser irradiance is increased, the trailing edge of the laser pulse will enhance the dissociation of C_n by multiphoton absorption through laser plasma interaction. This could cause the formation of C_2 nearer the target surface. Koren and Yeh [100] also have observed such double peaks in the case of C_2 from polymide target and their interpretation in terms of the fast and slow components does not completely explain our observations. We are therefore compelled to attribute the double peak formation due to the delays inherent in the distinctly different formation mechanisms of C_2 in the plasma. From fig. 10, it is seen that at higher spatial distance from the target (≥ 16 mm), the recombinational peak splits further into two forming a three fold TOF profile for the C_2 species. The reason for the occurrence of twin peak structure in the recombination emission intensity profile can be attributed to delays caused by different recombinational formation mechanism of C_2 species in the plasma.

From fig. 15, we see that the time delay for Pk1 increases with respect to increase in laser irradiance which is against the normal observation where velocity usually increases with the intensity of the incident laser pulse. By considering the velocity distribution, this anomaly can be explained only if one can have some type of a "negative diffusion" or anomalous diffusion of C_2 towards the target. Similar observations by one of the previous workers have been explained as due to selective depletion of high velocity C_2 species [51]. As the laser irradiance increases, the electron, atom and ion number densities of the plasma and plasma temperature are also increased. This may cause larger probability for events (24) and (25) indicated in the previous paragraph. It is also believed that ions are accelerated by Coulomb fields generated by fast moving electrons escaping from the plume. One should also consider the multiphoton dissociation of C_n by the trailing edge of the laser pulse so as to produce C_2 nearer to the target surface. Due to these processes C_2 may thus be formed at points nearer to the target surface by recombination processes, but with more than usual delay in the appearance of Pk1. In other words, decrease in delay with laser irradiance compete with the increase in delay due to shift of "formation site" of C_2 nearer to the target surface. Therefore, as far as Pk1 is concerned, delay due to the location of formation site is more predominant in the range of laser irradiances considered here thereby causing the observed enhancement in the delay on increasing the laser irradiance. This is further born out by the fact that there is a perceptible increase in the half width of peak Pk1 as the intensity of the laser pulse is increased.

Similar arguments can be made in the case of the measured time delays corresponding to Pk2. Contrary to the case of Pk1, delay of Pk2 remains almost constant up to a threshold laser irradiance and thereafter it decreases. As noted earlier Pk2 is found to be formed due to dissociation of carbon clusters C_n $(n > 2)$. As the laser irradiance is increased the formation of C_2 through dissociation of C_n will occur nearer to the target surface, thereby causing an increase in the delay as opposed to its decrease due to enhancement of kinetic energy of the species with laser irradiance. Up to a threshold laser irradiance, these two time delays are more or less balanced to give an effectively constant time delay as observed in fig.16 Above this threshold irradiance, enhancement in kinetic energy becomes so large that there will be a notable decrease in time delay for Pk2.

In order to understand the mechanism more clearly, experiments were carried out by varying the helium gas pressure in the plasma chamber. The pressure of the ambient atmosphere is one of the controlling parameters of the plasma characteristics, as well as factors related to the laser energy absorption. Variation of time delay for the peaks Pk1 and Pk2 with respect to helium ambient gas pressure is shown in figs. 17 and 18 respectively. It is interesting to note that the delay for PK1 decreases and for Pk2 it is increased with increasing helium pressure. The presence of helium gas helps cooling of plasma so as to reduce the plasma temperature. It is also supported by the fact that the increase of helium pressure decreases vibrational temperature of C_2 species [48].

The TOF profile of C_2 is found to be very sensitive to the pressure of the helium gas (fig. 19). With increasing pressure of the helium gas the temporal profile of Pk1 becomes more and more narrow, while for Pk2, it broadens. It is also noted that the intensity of the Pk1 increases steadily with increase in helium pressure. But in the case of Pk2, the intensities are almost constant at low pressures, then it enhances. The helium serves to cool the hot electrons by collisions leading to a more efficient electron impact excitation and plasma recombination, leading also to the plasma confinement near to the target which enhances the emission from these species.

The presence of helium gas during the ablation process has dramatic consequences on the expansion dynamics. The present results show that laser irradiance and pressure of the ambient helium gas have opposite effects on the plasma expansion processes. As irradiance increases, time delay for Pk1 increases while time delay for Pk2 is more or less constant up to a threshold irradiance and then decreases. Similarly increase in helium gas pressure decreases the time delay for Pk1 and in the case of Pk2, time delay is being constant up to a particular pressure and then increases. Pressure and laser irradiance also showed opposite effects on narrowing and broadening of Pk1 and Pk2. With increasing irradiance, a broadening and narrowing phenomena is observed for Pk1 and Pk2 respectively, just opposite to the case observed with increasing ambient gas pressure. These interesting results can be explained by assuming a simple adiabatic model. The plume length L can be estimated if it is assumed that at the end of the laser pulse the ablation products occupy a volume V_i at high pressure and subsequently expand adiabatically. Assuming adiabatic expansion model [108, 109], the ablated material pushes the gas species until the plasma and gas pressures equilibrate. Then the length of the plasma is given by

$$L = A[(\gamma - 1)E]^{1/3\gamma} P_0^{-1/3\gamma} V_i^{(\gamma-1)/3\gamma} \tag{25}$$

where A is the geometric factor related to the shape of the laser spot at the target surface, γ is the specific heat ratio (C_p/C_v), E is the laser energy density, P_0 is ambient gas pressure, V_i is the initial volume of the plasma ($V_i = v_0 \tau_{laser} w$, v_0 being the initial species velocity, τ_{laser} is the laser pulse duration and w, the laser spot size at the target surface). The factor A depends on the expansion geometry and for a conical plume with a spherical tip (inset of fig. 20) and expansion angle θ (for circular laser spot)

$$A = (1 + 1/\tan\theta)\left(\frac{3\tan\theta}{\pi + 2\pi\tan\theta}\right)^{1/3} \tag{26}$$

From eqn. 26, it is clear that E/P_0 is the parameter controlling the length of the plasma if the experimental geometry remains constant; therefore, E and P_0 have opposite effects as experimentally observed in the present work. According to this model, the length of the plasma L is expected to increase as the laser energy density is increased or ambient gas pressure is decreased. Considering the experimental

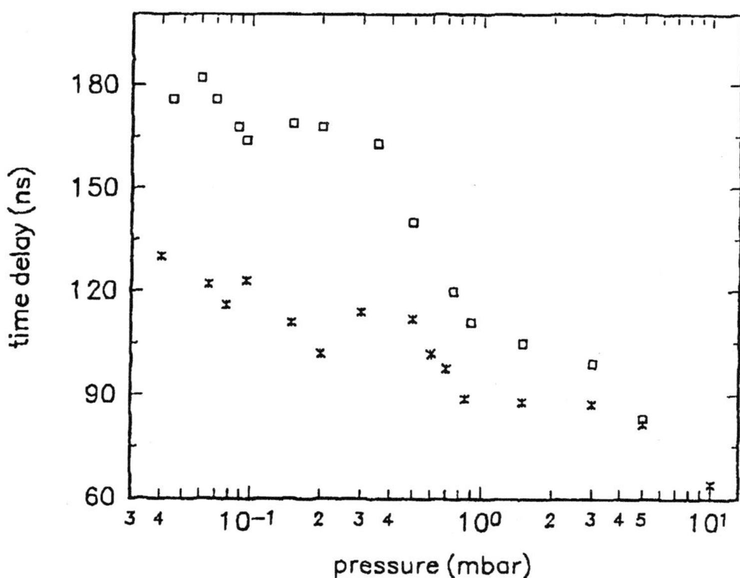

Figure 17: Dependence of time delays of Pk1 on ambient helium pressure ($\star$) 5 mm and (box) 8 mm (laser irradiance 81.34 GW cm^{-2})

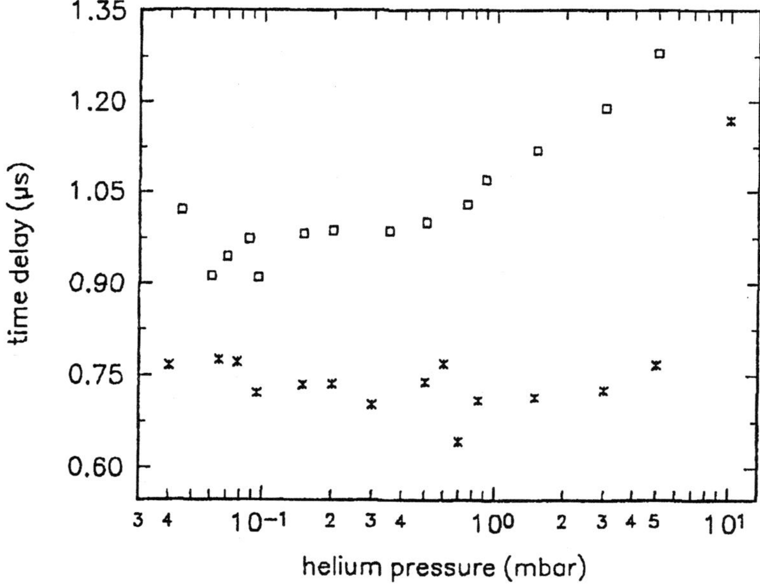

Figure 18: Dependence of time delays of Pk2 peak on ambient helium pressure ($\star$) 5 mm and (box) 8 mm (laser irradiance 81.34 GW cm^{-2})

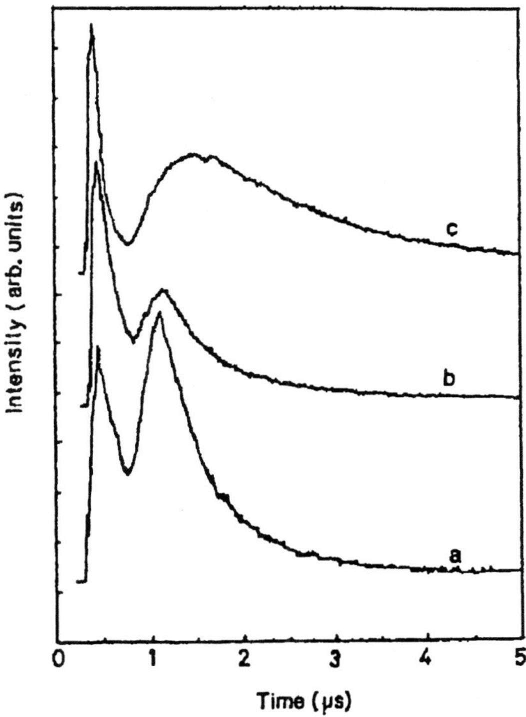

Figure 19: Intensity variation of spectral emission with time for C_2 observed at 8 mm away from the target at different ambient helium pressures (laser irradiance 81.34 GW cm^{-2}) (a) 0.065 mbar, (b) 1 mbar and (c) 10 mbar

parameters used in this work, evaluating eqn. 26 with $\theta = 35^0$ (then A $= 1.6$), $\gamma = 2.56$ [21], $v_0 = 4\text{-}8 \times 10^6 cms^{-1}$, $\tau_{laser} \simeq 9$ ns, gives a straight line as shown in fig. 20 corresponding to the present experimental pressure range. In our experiments, the estimated length of the plasma at a helium pressure of 0.05 mbar is 24 mm. Both the predicted pressure dependence and absolute value of L are seen to be reasonably in good agreement with experiment. The small error occurring in these calculations is mainly due to uncertainties in the value of γ and θ. It must be noted that the deposition rate and kinetic energy of ablated species will fall into a very low value for substrates located beyond L because of the subsequent cooling which tends to pull the material back to the target [110].

From fig. 13, one sees an anomalous variation of expansion velocity of the C_2 species with axial separation from the target. Gas dynamic effects are thought to play a major role in determining the spatial and velocity distribution of the vaporized material. The velocity at any point inside the plasma depends upon the spatial separation from the target surface. The atoms, molecules and ions undergo collisions in the high density region near the target forming the so called Knudsen layer, to create a highly directional expansion perpendicular to the target [19]. The density of the plasma will be maximum near the target due to collisions and hence the mean drift velocity normal to surface will be a minimum at very close to the target. The plasma expansion in a direction perpendicular to the target surface can be written as [28]

$$Z(t) \left[\frac{1}{2t} \frac{dZ}{dt} + \frac{d^2Z}{dt^2} \right] = \frac{6kT_0}{M} \tag{27}$$

where dZ/dt is the expansion velocity of the plasma in the Z direction, k, the Boltzmann constant, T_0 the isothermal temperature of the plasma and M is the molecular weight of the particle. The above equation evidently shows that the expansion velocities are low and acceleration is very high during the initial stages of expansion. But when the expansion velocities increase, acceleration starts to decrease. The laser pulse being short, the plasma cloud within the pulse duration has a minimum size in the direction normal to the target surface. A dense plasma can absorb strongly the trailing part of the laser pulse. Thus the absorption of the laser radiation by the plasma increases the velocity of the species inside the plasma and the hydrodynamic expansion is directed right angles to the surface of the target. This is found to be true as seen in figs. 13 and 14 respectively at distances near the target where expansion velocities increase very rapidly with respect to the spatial separation from the target. During the initial stages of the plasma expansion, the velocity in the direction perpendicular to the target surface is very high as a result of the small dimension of the plasma in that direction. It is reported that [111, 112] during shorter time intervals the expansion of the plume is one dimensional while for longer time scales the expansion is essentially three dimensional. This has also been supported by ultrafast photography of laser ablation plumes [113]. The dimensionality of the laser generated plasma, may deviate from these values

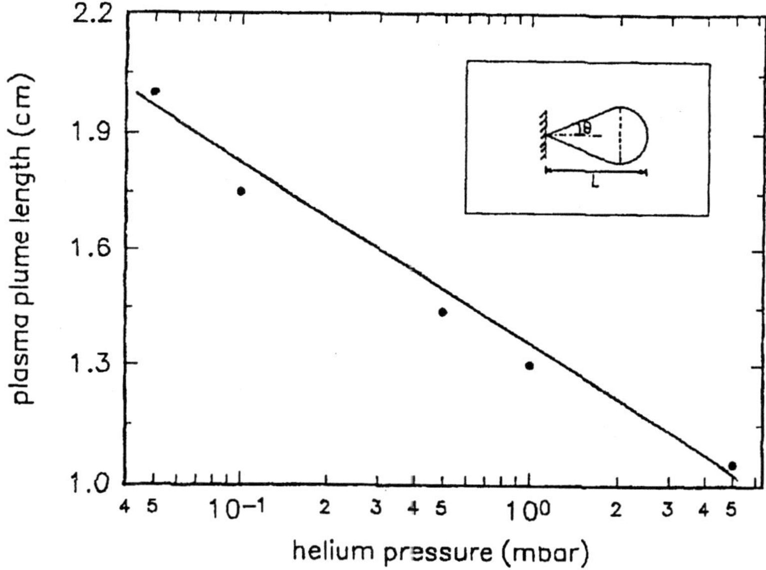

Figure 20: Estimated length of the ablation plume L as a function of helium pressure, laser irradiance 81.34 GW cm^{-2}
(inset shows plasma expansion geometry)

depending several factors including laser irradiance, plasma temperature, irradiated spot shape, mass of the species etc.

After the termination of the laser pulse, adiabatic expansion of the plasma begins. During this process, the thermal energy is converted to kinetic energy with the plasma attaining high expansion velocities. The adiabatic expansion of the plasma in the Z direction can be written as [28]

$$Z(t)\left[\frac{d^2Z}{dt^2}\right] = \frac{6kT_0}{M}\left[\frac{Z_0}{Z(t)}\right]^{\alpha-1} \qquad (28)$$

where Z_0 is the edge of the plasma at which the laser pulse is terminated and α, the maximum attainable particle velocity. In the adiabatic expansion region, the acceleration depends on the initial temperature and the mass of the species.

From fig. 14 it is seen that, the velocity of Pk2 is constant (8 km/s) after the initial expansion of the plasma. The velocities of these are found to be decreased at the boundary of the plasma which is in accordance with drag model [3] which predicts that the plume eventually comes to rest due to resistance from collision with background gas. For low pressures, when the driven mass of the background gas is small compared with the driver mass of the expanding vapour, the drag force model can be applied. In this case the viscous force is proportional to the velocity of ejected material and the plasma propagation can thus be written as

$$Z(t) = Z_0\left[1 - exp\frac{u_0 t}{Z_0}\right] \qquad (29)$$

where u_0 is the initial expansion velocity and $Z_0 = u_0/\beta$ the so-called stopping distance.

Fig. 13 shows that between distances 10 mm to 15 mm from the target surface, the expansion velocity decreases for Pk1 while it increases again at farther distances. This peculiar velocity pulsations can be explained as follows. During the adiabatic expansions, the thermal energy is rapidly converted into kinetic energy, thereby attaining high expansion velocities. It has been reported that for spherical plasmas, temperature drops off as the square of its radius [114]. A rapid drop in temperature occurs when the spherical plasmas expand. This may be the reason for the decrease in kinetic energy for Pk1 in the region between 10 mm to 15 mm. However the temperature drop will not continue with respect to square of the radius of the spherical plasmas indefinitely, because cooling due to expansion will be balanced by the energy gained from the recombination processes of the ions.

5   Conclusions

1.06 μm radiation from a Q-switched Nd:YAG laser is focused onto the carbon target where it produces a transient and elongated plasma. The dependence of electron density and electron temperature on different experimental parameters like distance from the target surface, time after the initiation of plasma, laser irradiance and pressure and nature of ambient gas used are carried out. At very short distances and times the intense continuum radiation is emitted from the target surface. The line to continuum ratio improves as time evolves. An initial electron temperature of about 3.6 eV is observed and it decays rapidly to much lower values within 200 ns time. At greater times the electron temperature is more or less constant. The electron density exhibits an approximate $1/z$ dependence with spatial separation from the target surface.

With increase in laser irradiance, both electron temperature and density increase and saturate at higher irradiance levels. Saturation phenomena of these fundamental parameters with laser irradiance can be explained by assuming the formation of self-regulating regime in the plume. The electron temperature and density show an abrupt change with the addition of ambient gases and these parameters also depend on the nature and composition of the gas used. It is noted that hotter and denser plasmas are formed in Ar atmosphere compared to He and air as a result of difference in the efficiency of cascade-like growth of electron number density and plasma absorption coefficient. Helium, being the lighter gas compared to Ar or N_2, gives rise to rapid cooling. Because of this, helium gas is preferred as a buffer instead of air or argon for the preparation of fullerenes since the enhanced clustering phenomena is more favoured in helium ambient. However, contrary to helium and argon ambient, the change in temperature profile in air, particularly at higher pressures, suggest that T_e may be affected by chemical reactions as well as simple hydrodynamic expansion of the plasma.

TOF studies of various species has differentiated the various mechanisms of the formation of C_2 species in the laser produced plasma from graphite in a helium gas atmosphere. Measurements of spatial dependence of the TOF emission intensities are made up to 25 mm away from the target. At low laser irradiances only a slowly propagating component of C_2 species is seen. At high irradiances the TOF profile shows a twin peak distribution. At distances greater than 16 mm from the target surface, a three fold TOF distribution is observed. The peak due to low kinetic energy component which is observed at all levels of irradiance is formed as a result of dissociation of higher clusters. The departure of single peak velocity distribution at higher laser irradiances is due to processes like recombination of the high energetic particles. The energy released on recombination is converted into kinetic energy of the molecules, atoms and ions, which may give rise to a group of relatively faster C_2 species. At farther distances from the target the recombinational peak gets modified in to two due to inherent delays caused by different recombination and excitation mechanisms. It is also found that these results are consistent with kinetic energy distribution of ionic species. The different expansion dynamics of C_2 species in the ambient gas are also discussed. The velocity pulsations in the faster peak during expansion into ambient gas are attributed to nonequilibrium kinetic energy transfer because of many body recombination.

The present results also show that the ambient gas pressure and laser irradiance have opposite effects on the TOF profiles of C_2 species. The broadening and narrowing of the multiple peaks with these parameters is expected due to collision and recombination phenomena. A simple adiabatic expansion model appears to provide a good description of the plume range and this may prove useful for scaling deposition experiments in terms of pressure and laser irradiance.

Acknowledgments: My thanks are due to Professor C. P. Girijavallabhan and Professor V. P. N. Nampoori, International School of Photonics, Cochin University of Science & Technology for fruitful discussions and support during the preparation of this manuscript. I am also grateful to Dr. C. V. Bindhu for her timely help.

References

[1] J. F. Ready, *Effects of High Power Laser Radiation* (Academic, London 1971).

[2] D. B. Chrisey and G. K. Hubler (eds.) *Pulsed laser deposition of thin films* (John Wiley and sons, New York, 1994).

[3] D. B. Geohegan, *Thin Solid Films* **220**, 138 (1992).

[4] D. B. Geohegan, A. A. Puretzky and D. J. Rader, *Appl. Phys. Lett.* **74**, 3788 (1999).

[5] E. A. Rohlfing, *J. Chem. Phys.* **93** 7851 (1990).

[6] S. M. Pimenov, V. V. Kononenko, V. G. Ralchenco, V. I. Konov, S. Gloor, W. Luthy, H. P. Weber and A. V. Khomich *App. Phy. A* **69** 81 (1999).

[7] S. S. Harilal, C. V. Bindhu, R. C. Issac, V. P. N. Nampoori and C. P. G. Vallabhan, *J. Appl. Phys.* **82** 2140 (1997).

[8] G. K. Varier, R. C. Issac, S. S. Harilal, C. V. Bindhu, V. P. N. Nampoori and C. P. G. Vallabhan, *Spectrochimica Acta B* **52** 1791 (1997).

[9] S. S. Harilal, R. C. Issac, C. V. Bindhu, V. P. N. Nampoori and C. P. G. Vallabhan, *J. Appl. Phys.* **81** 3637 (1997).

[10] S. S. Harilal, R. C. Issac, C. V. Bindhu, V. P. N. Nampoori and C. P. G. Vallabhan, *J. Appl. Phys.* **80** 3561 (1996).

[11] S. S. Harilal, R. C. Issac, C. V. Bindhu, V. P. N. Nampoori and C. P. G. Vallabhan, *Spectrochimi. Acta A* **53** 1527 (1997).

[12] S. S. Harilal, R. C. Issac, C. V. Bindhu, V. P. N. Nampoori and C. P. G. Vallabhan, *Pramana - J. Phys.* **46** 145 (1996).

[13] S. S. Harilal, P. Radhakrishnan, V. P. N. Nampoori and C. P. G. Vallabhan, *Appl. Phy. Letts.* **64** 3377 (1994).

[14] Z. Andreic, L. Aschke, H.-J. Kunze *J. Phys. D* **31** 1487 (1998).

[15] G. J. Pert *J. Plasma Phys.* **49** 295 (1993).

[16] R. F. Wood, K. R. Chen, J. N. Leboeuf, D. B. Geohegan and A. A. Puretzky *Phys. Rev. Lett.* **79**, 1571 (1997).

[17] R. F. Wood, J. N. Leboeuf, K. R. Chen, K. R. Chen, D. B. Geohegan and A. A. Puretzky *Appl. Surf. Sci.* **127-129**, 151 (1998).

[18] R. Kelly and A. Miotello in *Pulsed laser deposition of thin films* D. B. Chrisey and G. K. Hubler (eds.)(John Wiley and sons, New York, 1994) chap. 3.

[19] R. Kelly and R. W. Dreyfus, *Sur. Sci.* **198** 263 (1988).

[20] R. Kelly, *Phys. Rev. A* **46** 860 (1992).

[21] R. W. Dreyfus, R. Kelly and R. E. Walkup, *Nucl. Instru. Meth. Res.* B **23** 557 (1987).

[22] E. G. Gamaly, *Laser and Partcle Beams*, **12**, 185 (1994).

[23] E. G. Gamaly and R. Dragila *Phys. Rev. A*, **42**, 929 (1990).

[24] S. S. Harilal, C. V. Bindhu, V. P. N. Nampoori and C. P. G. Vallabhan, *Appl. Phys. Lett.* **72** 167 (1998).

[25] R. C. Issac, P. Gopinath, S. S. Harilal, V. P. N. Nampoori and C. P. G. Vallabhan, *Appl. Phys. A* **67** 557 (1998).

[26] S. S. Harilal, C. V. Bindhu, V. P. N. Nampoori and C. P. G. Vallabhan, *Appl. Phys. B* **66** 633 (1998).

[27] R. C. Issac, S. S. Harilal, C. V. Bindhu, V. P. N. Nampoori and C. P. G. Vallabhan, *Spectrochimi Acta B* **52** 1791 (1997).

[28] R. K. Singh, O. W. Holland and J. Narayan, *J. Appl. Phys.* **68** 233 (1990).

[29] R. K. Singh and J. Narayan, *Phys. Rev. B* **41** 8843 (1990).

[30] R. C. Issac, K. V. Pillai, S. S. Harilal, V. P. N. Nampoori and C. P. G. Vallabhan, *Appl. Surf. Sci.* **125** 227 (1998).

[31] H R. Griem, Plasma spectroscopy (McGraw-Hill Book Company, NewYork, 1964).

[32] H. R. Griem, *Spectral Line Broadening by Plasmas* (Academic Press, New York, 1974).

[33] H. R. Griem, Principles of Plasma spectroscopy (Cambridge, NewYork, 1997).

[34] R. H. Huddelestone and S. L. Leonard, *Plasma Diagnostic Techniques* (Academic press, London, 1965).

[35] D. H. Lowndes, D. B. Geohegan, A. A. Puretzky, D. P. Norton and C. M. Rouleau, *Science* **273** 898 (1996).

[36] I. H. Hutchinson, *Principles of Plasma diagnostics*, (Cambridge, NewYork, 1987).

[37] M. A. Heald and C. B. Wharton, *Plasma diagnostics with microwaves* (Wiley, New York, 1965).

[38] T. Mochizuki, K. Hirata, H. Ninomiya, K. Nakamura, K. Maeda, S. Horiguchi and Y. Fujiwara, *Opt. Commn.*, **72**, 302 (1989).

[39] S. B. Cameron, M. D. Tracy and J. F. Camacho, *IEEE Trans. Plas. Sci.*, **24**, 45 (1996).

[40] H.-J.Wesseling and B. Kronast *J. Phys. D* **29** 1035 (1996).

[41] J. M. de Regt, J. A. M Van de Mullen and D. C. Schram, *Phy. Rev. E*, **52**, 2982 (1995).

[42] F. E. Irons, *J. Phys. B*, **6**, 1562 (1973)

[43] H. R. Griem *Phys. Rev.*, **128**, 515 (1962)

[44] I. Ahmed, S. Buscher, Th. Wrubel and H.-J. Kunze *Phys. Rev. E*, **58**, 6524 (1998).

[45] H. W. Kroto, R. J. Heath, S. C. O'Brien, R. F. Curl and R. E. Smalley, *Nature* **318** 165 (1985).

[46] S. S. Harilal, C. V. Bindhu, V. P. N. Nampoori and C. P. G. Vallabhan, *J. Appl. Phys.* **86** 1388 (1999).

[47] S. S. Harilal, R. C. Issac, C. V. Bindhu, V. P. N. Nampoori and C. P. G. Vallabhan, *Pramana - J. Phys.* **49** 317 (1997).

[48] S. S. Harilal, R. C. Issac, C. V. Bindhu, V. P. N. Nampoori and C. P. G. Vallabhan, *J. Phys. D* **30** 1703 (1997).

[49] M. S. Donely and J. S. Zabinski, in *Pulsed laser deposition of thin films* D. B. Chrisey and G. K. Hubler (eds.)(John Wiley and sons, New York, 1994) Chapter 18.

[50] D. L. Pappas, K. L. Saenger, J. J. Cuomo and R. W. Dreytus, *J. Appl. Phys.* **72** 3966 (1992).

[51] D. L. Pappas, K. L. Saenger, J. Bruley, W. Krakow, J. J. Cuomo, T. *Gu* and R. W. Collins *J. Appl. Phys.* **71** 5675 (1992).

[52] C. B. Collins and F. Davanloo *in Pulsed Laser Deposition of Thin Films*, D. B. Chrisey and G. K. Hubler (eds) (John Wiley & Sons, New York, 1994).

[53] G. Dollinger, C. M. Frey and P. Maier-Komor, *Nucl. Instr. and Meth.* **A334** 167 (1993).

[54] S. Fujimori, T. Kasai and T. Inamura, *Thin Solid Films*, **92**, 71 (1982).

[55] A. A. Voevodin and M. S. Donley, *Surf. Coat. Technol.*, **82**, 199 (1996).

[56] W. R. Creasy and J. T. Brenna, *J. Chem. Phys.*, **92**, 2269 (1990).

[57] P. T. Murray and D. T. Peeler in *Laser Ablation : Mechanisms and Applications II* (American Institute of Physics, New York, 1994) p.359.

[58] F. Amiranoff, R. Fabbro, E. Fabre, C. Garban, J. Virmont and M. Weinfeld, *Phy. Rev. Letts.* **43** 522 (1979).

[59] J. Stevefelt and C. B. Collins, *J. Phys.* D **24** 2149 (1991).

[60] Y. Iida and E. S. Yeung, *Appl. Spectros.* **48** 945 (1994).

[61] A.P.Thorne, *Spectrophysics* (Chapman and Hall, London, 1974).

[62] G. Bekefi, *Principles of Laser Plasmas* (John Wiley & Sons, New York, 1976).

[63] W. L. Wiese, *Plasma Diagnostic Techniques*, R. H. Huddlestone and S. L. Leonard (eds), (Academic Press, New York, 1965)

[64] G. V. Marr, *Plasma Spectroscopy*, (Elsevier pub. company, New York, 1968).

[65] R. M. Herman, *Bull. Ame. Phys. Soc.*, **5** 234 (1960).

[66] S. S. Harilal, C. V. Bindhu, V. P. N. Nampoori and C. P. G. Vallabhan, *Appl. Spectroc.* **52** 449 (1998).

[67] S. S. Harilal, R. C. Issac, C. V. Bindhu, V. P. N. Nampoori and C. P. G. Vallabhan, *Plasma Sour. Sci. Tech.* **6** 317 (1997).

[68] G. Dinescu, E. Aldea, M. L. D. Giorgi, A. Luches, A. Perrone and A. Zocco, *Appl. Surf. Sci.*, **127-129** 697 (1998).

[69] J. Hermann, A. L. Thomann, C. Boulmer-Leborgne, B. Dubreuil, M. L. De Giorgi, A. Perrone, A. Luches and I. N. Mihailescu, *J. Appl. Phys.*, **77**, 2928 (1995).

[70] P. T. Rumsby and J. W. M. Paul, *Plasma Phys.*, **16**, 247 (1974).

[71] P. Mulser, R. Sigel and S. Witkowski, *Phys. Rep.*, **6**, 187 (1973).

[72] Ya. B. Zel'dovich and Yu. P. Reizer, *Physics of Shock Waves and High Temperature Dynamics Phenomena* (Academic, London 1968).

[73] J. J. Chang and B. E. Warner, *Appl. Phys. Lett.*, **69**, 473 (1996).

[74] A. Caruso and R. Gratton, *Plasma Phys.*, **10**, 867 (1968).

[75] L. Spitzer, *Physics of Fully Ionized Gases* (Wiley-Interscience, New York, 1962).

[76] S. S. Harilal, R. C. Issac, C. V. Bindhu, V. P. N. Nampoori and C. P. G. Vallabhan, *Jpn. Jour. Appl. Phys.* **36** 134 (1997).

[77] T. N. Hansen, J. Schoua and L. G. Lunney *Appl. Surf. Sci.* **138-139** 184 (1999).

[78] F. Garrelie and A. Catherinot, *Appl. Surf. Sci.* **138-139** 97 (1999).

[79] D. B. Geohegan, *Appl. Phys. Lett.*, **60**, 2732 (1992).

[80] R. C. Issac, P. Gopinath, C. P. G. Vallabhan and C. P. G. Vallabhan, *Appl. Phys. Lett.*, **73**, 163 (1998).

[81] R. F. Wood, J. N. Leboeuf, D. B. Geohegan and A. A. Puretzky and K. R. Chen,*Phys. Rev. B* **58**, 1533 (1998).

[82] A. L. Lewis and E. H. Piepmeier, *Appl. Spectrosc.*, **37**, 523 (1983).

[83] W. Sdorra and K. Niemax, *Spectrochimica Acta*, **45B**, 917 (1990).

[84] W. J. Treytz, K. W. Marich, J. B. Orenberg, P. W. Carr, D. C. Miller and D. Glick, *Anal. Chem.*, **43**, 1452 (1971).

[85] F. Leis, W. Sdorra, J. B. Ko and K. Neimar, *Mikrochimica Acta*, **II**, 185 (1989).

[86] Y. Iida, *Appl. Spectrosc.* **43**, 229 (1989).

[87] M. Ownes and V. Majidi, *Appl. Spectrosc.*, **45**, 1463 (1991).

[88] J. C. S. Kools, *J. Appl. Phys.*, **74**, 6401 (1993).

[89] D. B. Geohegan in *Laser Ablation : Mechanisms and Applications* J. C. Miller and R. F. Haglund (eds) (Springer-Verlag, Heidelberg 1991).

[90] D. B. Geohegan in *Pulsed Laser Deposition of Thin Films* D. B. Chrisey and G. K. Hubler (eds.)(John Wiley and sons, New York, 1994) chap. 5.

[91] S. S. Harilal, *Ph.d Thesis* Cochin Univ. of Sci. and Tech. (1997)

[92] M. Young and M. Hercher, *J. Appl. Phys.*, **38**, 4393 (1967).

[93] G. M. Weyl in *Laser-induced Plasmas and Applications*, L. J. Radziemski and D. A. Cremers (eds), (Marcel Dekker, Inc., New York, 1989) Chapter 1.

[94] R. G. Root in *Laser-induced Plasmas and Applications*, L. J. Radziemski and D. A. Cremers (eds), (Marcel Dekker, Inc., New York, 1989) Chapter 2.

[95] T. P. Hughes, *Plasmas and Laser Light*, (John Wiely and Sons, Inc., New York 1975).

[96] R. C. Weast, *CRC Handbook of Chemistry and Physics* (CRC press, Inc., Florida, 1988).

[97] Abhilasha, P. S. R. Prasad, and R. K. Thareja *Phys. Rev.* E **48** 2929 (1993).

[98] R. H. Dixon, R. C. Elton, and J. F. Seely, *Opt. Commun.*, **45** 397 (1983).

[99] R. H. Dixon and R. C. Elton, *Phys. Rev. Letts.*, **38** 1072 (1977).

[100] G. Koren and J. T. C. Yeh, *J. Appl. Phys.* **56** 2120 (1984).

[101] Y. Tasaka, M. Tanaka and S. Usami, *Jpn. J. Appl. Phys.* **34** 1673 (1995).

[102] Y. Tasaka, M. Tanaka and S. Usami, *Appl. Sur. Sci.* **79/80**, 141 (1994).

[103] A. V. Bulgakov and N. M. Bulgakova, *J. Phys.* D **28** 1710 (1995).

[104] K. A. Kummuduni, Y. Nakayama, Y. Nakata, T. Okada and M. Maeda, *J. Appl. Phys.* **74** 7510 (1993).

[105] A. Harano, J. Kinishita, K. Itou, T. Kitamori, T. Sawada and S. Koda, *J. Spectrosco. Soc. Japan,* **42** 94 (1993).

[106] H. W. Kroto, A. F. Allaf and S. P. Balm, *Chem. Rev.* **91** 1213 (1991).

[107] E. N. Sobol in *Phase Transformation and Ablation in Laser Treated Solids* (John Wiley and Sons, New York, 1994) chp.6.

[108] P. E. Dyer, A. Issa and P. H. Key, *Appl. Surf. Sci.* **46** 89 (1990).

[109] P. E. Dyer, A. Issa, P. H. Key, *Appl. Phys. Letts.* **57** 186 (1990).

[110] G. Koren, *Appl. Phys. Letts.* **51** 569 (1987).

[111] W. Pietsch, B. Dubreuil and A. Briand, *Appl. Phys.* B **61** 267 (1995).

[112] J. C.S. Kools, T. S. Baller, S. T. Dezwart and J. Dielemen, *J. Appl. Phys.* **71** 4547 (1992).

[113] A. Guptha, B. Baren, K. G. Casey, B. W. Husley and R. Kelly, *Appl. Phys. Letts.* **59** 1302 (1991).

[114] J. M. Dawson, *Phys. Fluids* **7** 981 (1964).

Index